BOY SCOUTS OF AMERICA
MERIT BADGE SERIES

AVIATION

Troop 3

BOY SCOUTS OF AMERICA®

Requirements

1. Do the following:
   a. Define "aircraft." Describe some kinds and uses of aircraft today. Explain the operation of piston, turboprop, and jet engines.
   b. Point out on a model airplane the forces that act on an airplane in flight.
   c. Explain how an airfoil generates lift, how the primary control surfaces (ailerons, elevators, and rudder) affect the airplane's attitude, and how a propeller produces thrust.
   d. Demonstrate how the control surfaces of an airplane are used for takeoff, straight climb, level turn, climbing turn, descending turn, straight descent, and landing.
   e. Explain the following: the recreational pilot and the private pilot certificates; the instrument rating.

2. Do TWO of the following:
   a. Take a flight in an aircraft, with your parent's permission. Record the date, place, type of aircraft, and duration of flight, and report on your impressions of the flight.
   b. Under supervision, perform a preflight inspection of a light airplane.
   c. Obtain and learn how to read an aeronautical chart. Measure a true course on the chart. Correct it for magnetic variation, compass deviation, and wind drift. Arrive at a compass heading.

d. Using one of many flight simulator software packages available for computers, "fly" the course and heading you established in requirement 2c or another course you have plotted.

e. On a map, mark a route for an imaginary airline trip to at least three different locations. Start from the commercial airport nearest your home. From timetables (obtained from agents or online from a computer, with your parent's permission), decide when you will get to and leave from all connecting points. Create an aviation flight plan and itinerary for each destination.

f. Explain the purposes and functions of the various instruments found in a typical single-engine aircraft: attitude indicator, heading indicator, altimeter, airspeed indicator, turn and bank indicator, vertical speed indicator, compass, navigation (GPS and VOR) and communication radios, tachometer, oil pressure gauge, and oil temperature gauge.

g. Create an original poster of an aircraft instrument panel. Include and identify the instruments and radios discussed in requirement 2f.

3. Do ONE of the following:
    a. Build and fly a fuel-driven or battery-powered electric model airplane. Describe safety rules for building and flying model airplanes. Tell safety rules for use of glue, paint, dope, plastics, fuel, and battery pack.
    b. Build a model FPG-9. Get others in your troop or patrol to make their own model, then organize a competition to test the precision of flight and landing of the models.

4. Do ONE of the following:
    a. Visit an airport. After the visit, report on how the facilities are used, how runways are numbered, and how runways are determined to be "active."
    b. Visit a Federal Aviation Administration facility—a control tower, terminal radar control facility, air route traffic control center, flight service station, or Flight Standards District Office. (Phone directory listings are under U.S. Government Offices, Transportation Department, Federal Aviation Administration. Call in advance.) Report on the operation and your impressions of the facility.
    c. Visit an aviation museum or attend an air show. Report on your impressions of the museum or show.

5. Find out about three career opportunities in aviation. Pick one and find out the education, training, and experience required for this profession. Discuss this with your counselor, and explain why this profession might interest you.

Contents

Introduction to Aviation. 7

How Airplanes Work. 14

All About Instrumentation. 33

Communication. 41

Air Navigation. 44

Entering the Cockpit. 56

Taking Flight. 59

Aviation Facilities. 69

Flying Without Leaving the Ground. 77

Careers in Aviation 91

Aviation Resources 94

Introduction to Aviation

Flying! For most of history, people have dreamed of flying, imagining how it would feel to soar through the sky like an eagle or hover in midair like a hummingbird, to float on unseen currents, free of Earth's constant tug, able to travel great distances and to rise above any obstacle.

Today, through aviation, we can not only join the birds but also fly farther, faster, and higher than they ever could. Today, aircraft routinely fly across the country, around the world, and even beyond Earth's atmosphere.

But our ability to fly is relatively new. It has been only in the last 100 years or so that people have mastered flight. Aviation's progress since then has been nearly as breathtaking as a ride in the latest Air Force jet.

The History of Flight

The first successful manned flight took place on November 21, 1783, in Paris, France—and it did not involve an airplane. That day, brothers Joseph and Etienne Montgolfier sent two men up in a hot-air balloon they had made out of cotton and paper. The men stayed aloft for 25 minutes and traveled about 5 miles.

From lighter-than-air balloons, early aviators progressed to the original heavier-than-air flying machines—gliders or sailplanes. In 1853, English engineer Sir George Cayley built the world's first real glider, which carried his terrified coachman across a small valley. Later in the century, German engineer Otto Lilienthal built a series of gliders in which he made regular, controlled flights.

Inspired by Lilienthal's gliders, two bicycle mechanics from Ohio, Orville and Wilbur Wright, began studying aviation and experimenting with their own aircraft. On December 17, 1903, they ushered in the aviation age when Orville took off from a sand dune near Kitty Hawk, North Carolina, traveled 120 feet in 12 seconds, and landed safely. Those 12 seconds changed history.

Charles Lindbergh and the *Spirit of St. Louis*

Aviation grew quickly in the decades after the Wright brothers' historic flight. In 1909, Glenn Curtiss made headlines for flying 142 miles from New York City to Albany, New York. In 1927, Charles Lindbergh made the first solo flight across the Atlantic Ocean. In 1938, Howard Hughes and his crew flew around the world in just under four days. And in 1969, Eagle Scout Neil Armstrong stepped onto the surface of the moon. Many people who watched Armstrong on television that day had heard about Lindbergh's flight on the radio and read about the Wright brothers in the newspaper.

U.S. Navy biplane, 1934

Types of Aircraft

The term **aircraft** is broad, covering nearly everything that enables people to fly through the air. Some aircraft (balloons, blimps) are lighter than air; others, like airplanes and helicopters, are heavier than air. (Missiles, rockets, and vehicles like the space shuttle are called **spacecraft** since they are designed to fly outside Earth's atmosphere.)

Here are some of the kinds of aircraft in use today. Each of these aircraft has been designed to do a particular job. You are probably already familiar with many of them.

- Commercial airliners
- Cargo airplanes (including those used by express delivery services)
- Military bombers, fighters, and surveillance aircraft
- Military supply transports
- Military and civilian helicopters and autogyros

Introduction to Aviation

- Personal-use and training airplanes of various types
- Airplanes owned by corporations to transport personnel
- Balloons for sport or for exploring the atmosphere
- Sailplanes, amphibians, and seaplanes
- Blimps used for advertising in the sky and as television camera platforms
- Aerobatic airplanes for exhibitions
- Crop sprayers
- Fire-fighting "smoke jumper" transports and "borate bombers"

Aircraft at Work

Even if you have never boarded an airplane, your life is affected by aircraft every day. Cargo planes carry many of the packages you receive in the mail. Helicopters make possible the traffic reports that identify trouble spots for commuters. Blimps provide bird's-eye views of sporting events. Military jets protect the skies above your home.

Other uses of aircraft may surprise you. Helicopters can be used to dry out the field before a baseball or football game. Hovering a few feet off the field, their giant rotors dry the surface with their powerful downwash of air. Construction companies use helicopters to "top off" tall buildings by lifting a structure's final pieces into place. Large transport planes serve as aerial pumper trucks, fighting forest fires.

INTRODUCTION TO AVIATION

Transportation. Most uses of aircraft are more familiar, such as the transportation of passengers. Aircraft make it possible for businesspeople to attend meetings in faraway cities and return home in time for dinner. They allow families to take vacations on the other side of the country without spending endless days in a car. Airplanes also allow people to conduct business internationally and visit other countries without taking long rides on a ship.

Geologists, surveyors, and forest rangers use helicopters and small airplanes to reach remote places. In some large cities, helicopters make regular flights from the congested business district to the airport, greatly reducing travel time. In rural areas, small airplanes sometimes serve as taxis in areas not served by major airlines and airports.

> Alaska has more airplanes per person than any other state; much of the state is easily accessible only by air.

Mail and Packages. Small airplanes have been used to carry mail since aviation's earliest days; the U.S. government started airmail service in 1918. Even today, winged carriers of mail and supplies may be the only regular face-to-face contact some people in remote areas have with the outside world.

The airplane is a key element in today's high-demand business world, where moving packages in as little time as possible is extremely important.

One nautical mile is equal to 6,076 feet.

The Military. Aircraft are among the U.S. military's most important tools, and air power continues to play a decisive role in conflicts around the globe. But aircraft do more than drop bombs on sites far behind enemy lines. They also carry out surveillance, serve as flying ambulances, and transport troops and equipment. The C-17 Globemaster III can carry 170,900 pounds of cargo and fly 2,800 nautical miles without refueling.

Firefighting and Public Safety. The USDA Forest Service spots forest fires from airplanes as well as from lookout towers. And once a fire is spotted, aircraft swing into action. Helicopters bring firefighters and equipment to remote areas and can extract them quickly when the need arises. Airplanes called **borate bombers** drop water or flame-retardant agents (which once included borate salts) on fires from above.

Helicopters have become indispensable in fighting wildfires, dropping water on hot spots to assist crews.

Helicopters are the unsung heroes of many rescues at sea and in rugged mountain regions. In the emergency medical field, paramedics use helicopters as air ambulances to quickly transport severely injured people from accident sites to hospitals.

Law Enforcement. The airplane's speed and maneuverability are valuable to many law enforcement agencies. Federal and state law enforcement agencies like the U.S. Department of Homeland Security, narcotics inspectors, and game wardens all use aircraft for patrol work, transporting officers, and chasing suspects.

Agriculture. Crop-spraying airplanes have been a common sight in the rural sky for many years. Today, helicopters are often used to spray crops because the "wash" from their rotors tends to distribute the spray more widely, even blowing it up to the underside of the crop's leaves.

Airplanes haul perishable items to market and help ranchers patrol fences, herd cattle, and bring feed to animals in distant pastures. Some farmers hire aviation companies for these services, but so much of the work is done by farmers themselves that they have formed an organization called the International Flying Farmers.

Aerial Photography. Aerial photography has been important for mapmakers and newsmen since the first intrepid cameraman leaned over the side of one of the early flying machines with his box camera. Now, much filming is done from helicopters because they can furnish a stationary platform in the sky for the camera operator.

How Airplanes Work

When people first dreamed of flying, they looked to the birds as models. Greek mythology includes the story of Daedelus and Icarus, who made wings out of wax and feathers to escape from the island of Crete. (Unfortunately, the story says, Icarus flew too close to the sun, which melted his wings.) Even as late as the 19th century, inventors were still trying to imitate birds with elaborate flying machines that never quite got off the ground.

Sir George Cayley had a better idea. His glider had fixed wings and a moveable tailplane. What it did not have was an engine—nobody had yet figured out how to make one that was light and powerful enough for aviation. Nonetheless, his ideas led more or less directly to the 1903 Wright Flyer, to the fighter planes of World War II, and to the massive airliners of today.

To understand how an airplane works, you need to learn about several different concepts: the forces that affect an airplane, what airfoils are and how they work, the control surfaces a pilot can use, and the types of engines that give an airplane power.

In 1488, Leonardo da Vinci came up with a plan for an "ornithopter," a winged flying machine like an oversized bird, though he never actually built one.

How Airplanes Work

Forces That Affect Airplanes

Gravity makes anything that has weight—even a feather—fall to Earth. So how can we get a 400-ton airplane like a Boeing 747 into the air and keep it there? By matching two natural forces with two forces that are man-made.

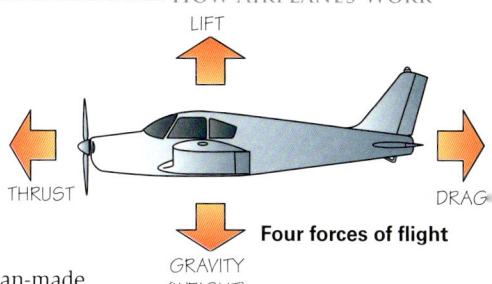

Four forces of flight

The natural forces are **gravity** and **drag.** Toss a ball into the air, and you will see gravity in action. Thanks to this downward-acting force, the ball quickly falls to the ground. Drag, on the other hand, is a backward-acting force. It is the resistance you feel when you pedal a bicycle into a strong headwind. Airplane designers offset these natural forces with two forces of their own: **thrust** and **lift.** These forces work together to pull an airplane forward and to keep it off the ground.

Lift is the upward-acting force; it is created by the way airplane wings are designed. Rather than slip easily past the wings, air pushes them upward. Thrust is the force that pushes the airplane forward; it is produced by the plane's powered propeller or jet engine.

Although aviation includes all sorts of manned aircraft, we will mostly focus on airplanes in this pamphlet.

Airfoils

Look at a wing from the side of an airplane, and you will notice two things. First, the bottom surface of the wing is more or less flat, while the upper surface curves upward. Second, the wing is attached at a slight angle instead of being level with the ground. Together, these two characteristics produce lift. They also make the wing an **airfoil,** which is a shaped structure that provides lift when air is forced across its surface.

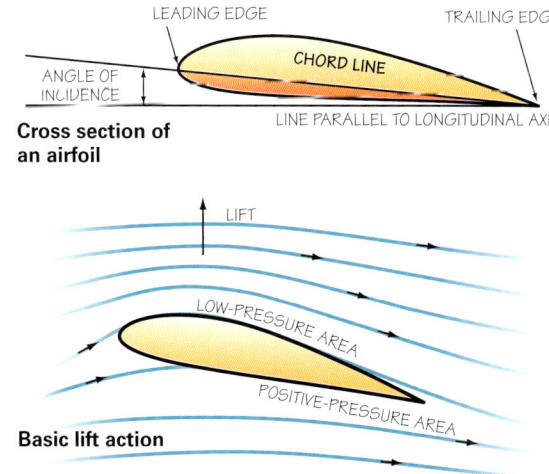

AVIATION 15

How Airplanes Work

The wings are not the only airfoils on an airplane—so are propeller blades. They work by creating greater air pressure on one side of their surfaces than on the other. As the blades cut through the air, they pull the plane along because the pressure behind them is greater than the pressure in front.

If you have ever stuck your hand out the window of a moving car, you have created a simple airfoil. Angle your palm upward and the wind lifts your hand. Flatten your hand out and the lift decreases. Just think: If your hands were big enough (and your arms strong enough), you could make your car fly.

In the 18th century, a Swiss scientist named Daniel Bernoulli discovered something about how all fluids (liquid or gas) work that is vital to flying a heavier-than-air machine. He found that a fluid's pressure decreases at points where its speed increases. For example, if you direct a stream of air into a pipe that is smaller in diameter in some places than in others, the pressure will be lowest and the speed fastest in the most confined section of the pipe.

> You can see Bernoulli's principle in action at your kitchen sink. Hold a teaspoon upside down and direct a stream of water at the back side of the bowl. You might expect the water to push the spoon down forcefully, but it actually hugs the bowl's contours, reducing the pressure on the bowl. Flip the spoon over, and you will get a completely different effect.

As an airfoil moves through the atmosphere, air must go faster to get over the curved top than to go under the flatter bottom. According to Bernoulli's principle, therefore, the air pressure is lower above the wing than under it; that creates lift.

The angle at which the wing is attached also creates lift. You have probably heard of Isaac Newton's third law of motion—for every action there is an equal and opposite reaction. When air hits the bottom of the wing, it is diverted down toward the ground. The reaction to that action is that the wing is pushed up toward the sky.

= HOW AIRPLANES WORK

Airfoils in Action

The angle at which a wing is attached is called the **angle of incidence.** As we have discussed, that angle creates lift, but it also produces greater drag, as you can see in the accompanying illustrations.

As a pilot flies the plane, he moves the entire aircraft and changes the angle at which the wings meet the air to maneuver the plane upward or downward. This angle is called the **angle of attack.** The greater this angle—up to a certain point—the more lift and drag there will be.

When the angle of attack reaches about 16 to 20 degrees, the air cannot flow smoothly over the wing's upper surface. This extreme angle makes the air churn behind the wing as it tries to follow the surface. There is a sudden increase in pressure on the upper wing surface. This immediately reduces lift and increases drag, and the wing is said to "stall." When this occurs, the plane does not have enough thrust or lift to hold it up. The plane can go out of control until the pilot reduces the angle of attack.

Lift and drag are also affected by several other factors such as the size of the wing, the shape of the airfoil, the speed of the airplane, and the density of the air itself. Let's examine the density of the air.

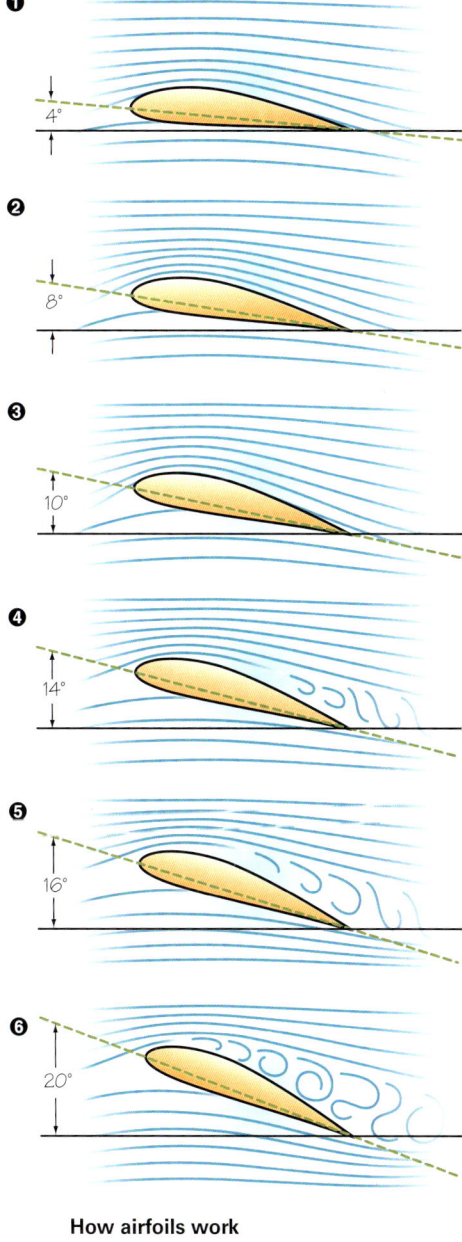

How airfoils work

AVIATION 17

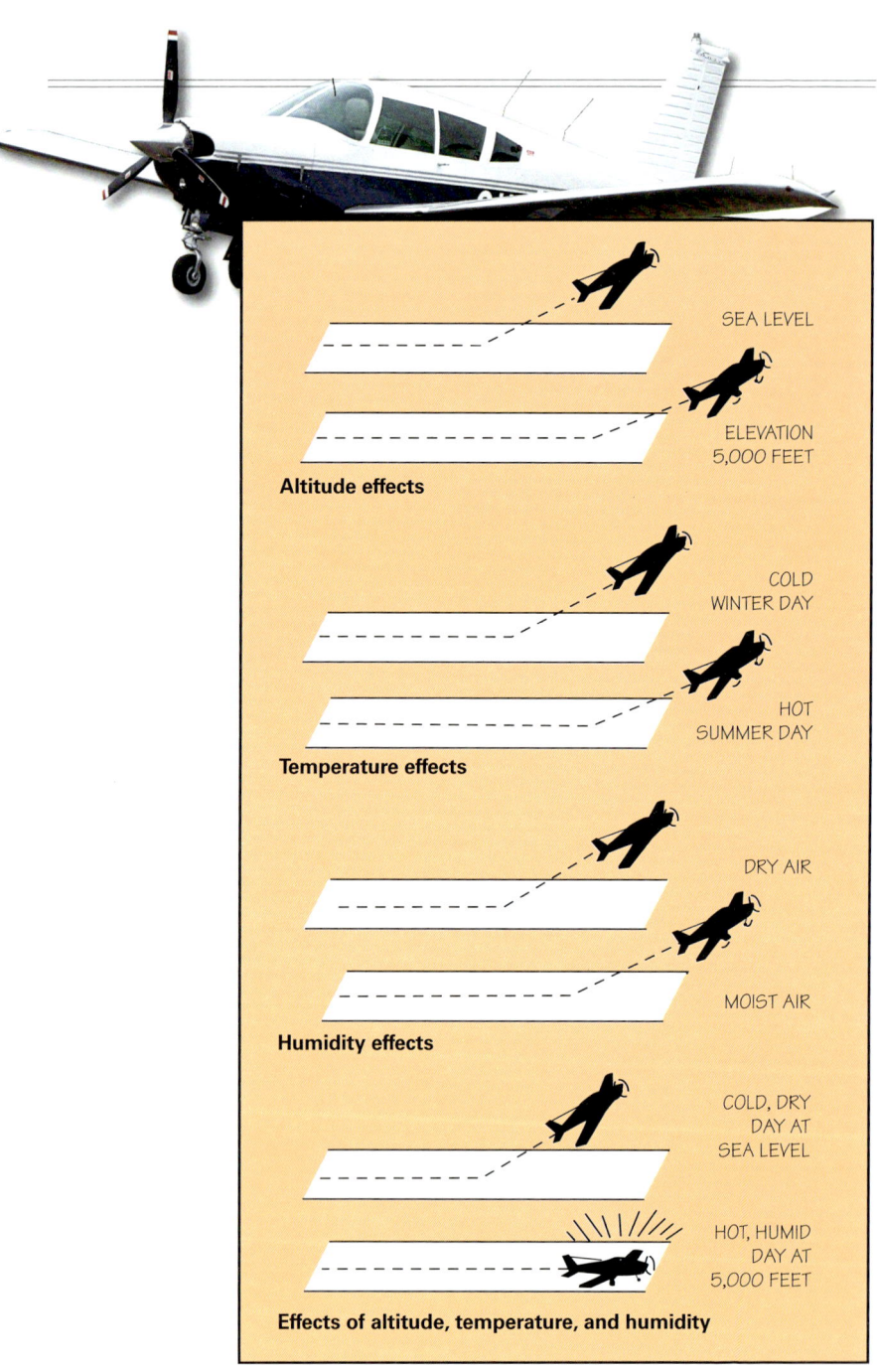

Air: The Pilot's Element

Air has weight and substance, so it tends to resist anything passing through it. We have seen how an airfoil lifts a plane by changing the air pressure (or weight) around it. But the air is constantly changing. Sometimes it is quite dense and heavy; at other times (and always at high altitudes) it is thin and light. Pilots must understand how and why the air changes so that they know what to expect.

How Density Changes

Three things affect the density of the air: altitude, temperature, and humidity. As the illustration shows, these factors can drastically change how an airplane functions, including how long a run it must take before getting aloft.

Our atmosphere extends hundreds of miles out from Earth. The farther from Earth you go, the less atmospheric pressure there is. At sea level, the atmosphere exerts about 20 tons of pressure on your body. (Fortunately, your body exerts an equal force on the atmosphere, keeping you from being crushed.) When your altitude reaches 18,000 feet, atmospheric pressure drops by half.

The second factor that affects air density is temperature. When air is heated, it expands and thus has less density. That means a pilot needs a longer runway to take off on a hot day than on a cool day.

The final factor is humidity—the amount of water vapor in the air. You might think that humid air would be more dense than dry air, but the reverse is actually true. Water vapor weighs less than perfectly dry air, so the air is denser on dry days than when the humidity is high.

To sum up the effects of temperature and humidity, we can say that a pilot will have to make a much longer run to take off on a warm, humid day than on a cool, dry day because the air will be less dense. If the airport is high above sea level, the takeoff will be harder still because the density of the air is lower.

These three factors—altitude, temperature, and humidity—are interrelated. Changes in temperature will change the humidity, and changes in altitude will affect both. Changes in pressure and temperature are largely responsible for the weather conditions—winds, storms, snow, and ice—that a pilot encounters on the ground and aloft.

An airplane that requires 1,000 feet of runway to take off at a sea-level airport will need almost 2,000 feet at Denver International Airport, which is a mile above sea level.

Building an Airfoil

When airplane designers create wings, they must consider two key factors:

1. The lift and drag acting on a wing are related to the size of the wing. That is, if two wings have the same shape but one is twice as large as the other, the bigger one will have twice the lift and twice the drag.

2. As the upper curve, or camber, of a wing is increased (up to a certain point), the lift produced by the wing increases. This is because the air must travel farther and faster around a large curve, thus reducing air pressure above the wing.

In many modern jet aircraft, the wings are rather short and stubby, and they may be quite thin. The speed of these jets is so great that a small airfoil can provide sufficient lift. Also, the small wings reduce drag. The space shuttle has delta wings, which are triangular and sweep back at a sharp angle from the fuselage. Some military airplanes also use this type of wing.

= How Airplanes Work

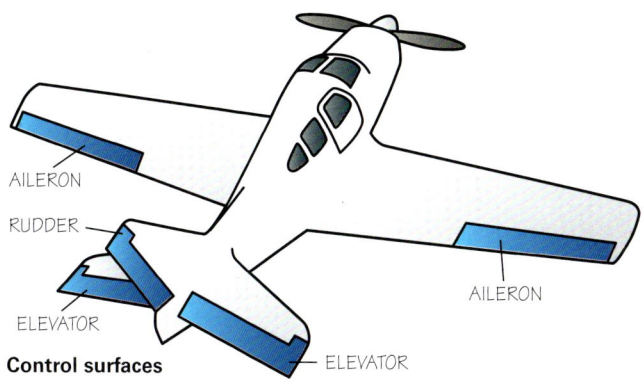

Control surfaces

Control Surfaces

Earlier we saw how a pilot can maneuver the entire airplane to change the angle of attack and increase or decrease lift. By manipulating the plane's **control surfaces,** the pilot can do even more, including steering, climbing, and descending.

Control surfaces are sections of five airfoils—the right and left wings, the two smaller areas on the tail called the horizontal stabilizer, and the vertical stabilizer, or fin. When the pilot moves these surfaces, the flow of air changes, affecting the plane's altitude and direction. The control surfaces on the wings are called **ailerons,** the control surfaces on the horizontal stabilizer are called **elevators,** and the control surface on the vertical stabilizer is called the **rudder.**

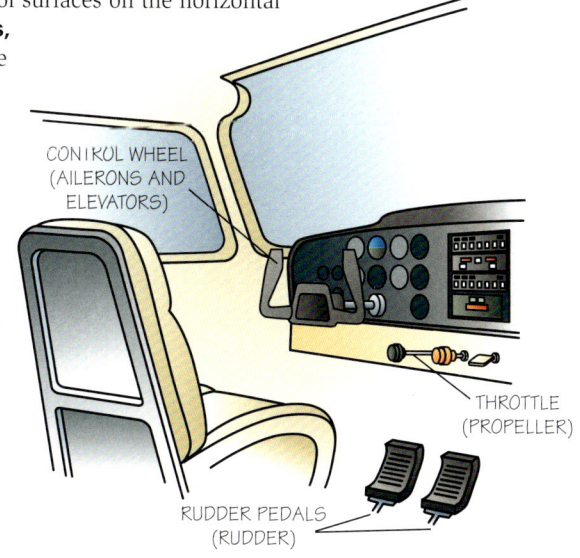

The pilot can control the propeller and control surfaces by using the control wheel, throttle, and rudder pedals.

AVIATION

How Airplanes Work

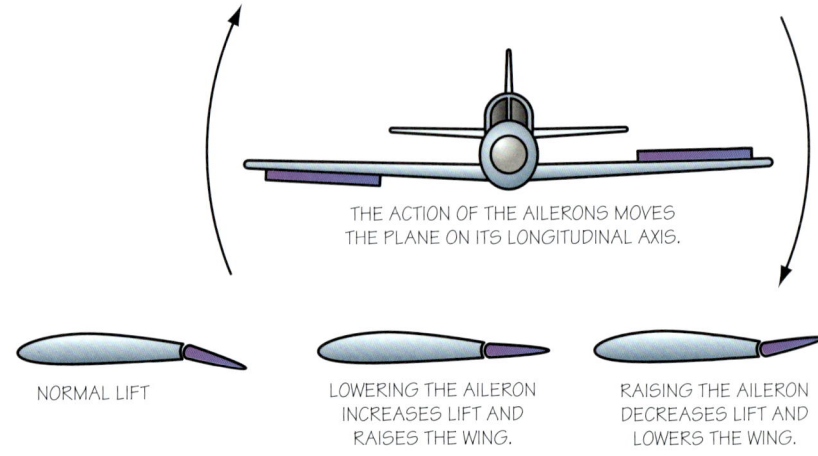

THE ACTION OF THE AILERONS MOVES THE PLANE ON ITS LONGITUDINAL AXIS.

NORMAL LIFT

LOWERING THE AILERON INCREASES LIFT AND RAISES THE WING.

RAISING THE AILERON DECREASES LIFT AND LOWERS THE WING.

The Control Wheel

The **control wheel** (also called the stick or yoke) can be moved in all directions. As it is moved, it changes the positions of the ailerons and elevators. When the wheel is rotated to the right, the left aileron goes down and the right one goes up. The plane rolls to the right, because these movements of the ailerons change the curvature of the wings' surfaces. Just the opposite happens when the pilot rotates the wheel to the left.

When the airplane's wings are not level, it is in a bank. Banking is how the pilot turns the airplane.

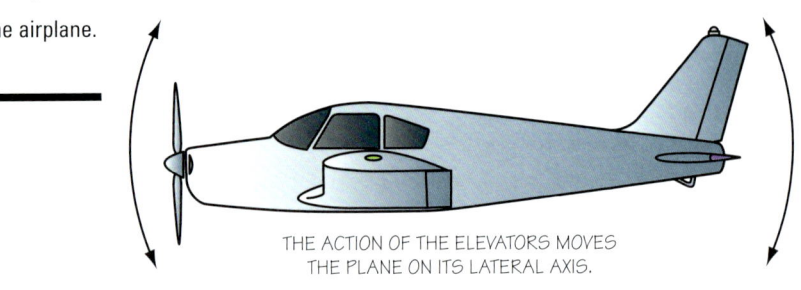

THE ACTION OF THE ELEVATORS MOVES THE PLANE ON ITS LATERAL AXIS.

RAISING THE ELEVATOR FORCES THE TAIL DOWN, AND THE NOSE RISES.

NORMAL LIFT

LOWERING THE ELEVATOR FORCES THE TAIL UP, AND THE NOSE DROPS.

AVIATION

If the pilot pushes the control wheel forward, the elevators move downward. This increases the lift provided by the horizontal tail assembly, forcing the tail up and the nose down. If the pilot pulls back on the wheel, the elevators turn upward, decreasing the tail assembly's lift. This forces the tail down and the nose up.

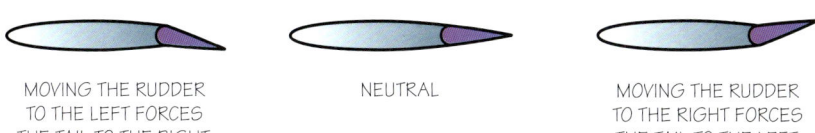

MOVING THE RUDDER TO THE LEFT FORCES THE TAIL TO THE RIGHT.

NEUTRAL

MOVING THE RUDDER TO THE RIGHT FORCES THE TAIL TO THE LEFT.

The rudder helps keep the plane flying straight by controlling what is called **yaw,** or side-to-side motion of the plane's nose. Rudder pedals in the cockpit connect to the rudder. If the pilot presses the right rudder pedal, the nose moves to the right. Pressing the left rudder pedal causes the nose to move to the left.

The **throttle,** which adjusts the engine's power, is an important factor in maneuvering. For example, you would not climb just by pulling back on the wheel to lower the plane's tail and raise the nose. Doing so would rapidly increase drag and cause the wing to lose considerable lift. What you would actually do is pull back on the wheel and, at the same time, advance the throttle to increase thrust and maintain airspeed. This would keep lift high and help the plane climb.

RUDDER

> Requirement 1c asks you to show how pilots use these control surfaces to maneuver a plane. You can demonstrate this by actually working a plane's controls on the ground if one is available to you. If it is not, you may demonstrate by using a model plane or by showing the positions with hand motions.

Maneuvering the Plane

Now, let's look at how pilots use a plane's control surfaces to handle several basic maneuvers.

Takeoff

When maneuvering a plane, the pilot applies and releases pressure using the controls, and the plane responds accordingly.

Takeoff (given an airplane with a tricycle landing gear). The elevators are neutral, ailerons level, and rudder centered. Use the rudder to steer the airplane down the runway. When flying speed is reached, pull back on the wheel to lift the nose. (The term pilots use is **rotate.**)

Straight climb. Control what is called the pitch attitude (the relationship between the nose position and a level attitude) with the elevators. On a typical airplane, the nose will be from 8 to 10 degrees above level for a normal climb.

Level turn. Turns are made by the ailerons and rudder working together. During a level turn, you must pull back some on the control wheel to maintain the airplane's altitude. To make a level turn to the left, turn the wheel to the left and press the left rudder pedal. The left aileron goes up, the right one goes down, and the rudder goes to the left. This forces the plane into a coordinated banking left turn in level flight.

How Airplanes Work

Banked turn

Banked turn. In a banked turn, the airplane lift is banked as well. Thus, the lift does not directly oppose the airplane's weight as it does when the wings are level. You must then raise the nose of the airplane (which increases the wing's angle of attack) to increase the lift. This will cause the aircraft to climb around the turn; stop pulling on the control wheel when rolling out of the turn, or the airplane will start to climb.

After establishing the desired degree of the bank, return the controls to neutral position. The plane will then continue to turn until you move the controls in the opposite direction of the turn to return the plane to straight and level flight.

Climbing turn. To climb, pull back on the wheel, forcing the plane's tail down and the nose up. So, to make a climbing turn, combine this movement with those of turning. For a climbing left turn, ease the control wheel back and to the left and press the left rudder pedal. At that point, the left aileron will be up (helping to lower the wing), the right aileron down (helping to raise the wing), the rudder to the left, and the elevators up. Neutralize the ailerons and rudder when you reach the desired angle of bank.

Descending turn. The position of the control surfaces for a descending turn is the same as for any other turn, except that you move the wheel forward slightly to lower the nose in relation to level and reduce the power unless you want to build up speed.

Straight descent. To descend on a straight course, make all control surfaces level except for the elevators. These are down slightly, which means the plane's tail is up and the nose down.

Landing

Landing. As the plane nears the ground, reduce power and pull back on the control wheel to slow the airplane. When the wheels are just above the ground, move the control wheel back, further decreasing the speed. The plane lands on the two main wheels.

The Science of Flight

Also located on the airplane's wings are **flaps, slats,** and **spoilers.** Manipulating these devices allows the pilot to increase or decrease the lift and drag by changing the shape of the wing. They are employed during takeoff and landing to help stabilize the aircraft's speed and direction.

Sources of Power

Without an engine, an airplane would be nothing but a glider. In fact, many aviation pioneers—including the Wright brothers—mastered glider flight before they began experimenting with engines. The goal was to build an engine with the right power-to-weight ratio. In other words, the engine had to be both powerful enough and light enough to make flight possible.

The Wright brothers used a piston engine they built themselves. The piston engine, which works much like the engine in your family's car, remained the primary source of airplane power for four decades. Soon after World War II, however, jet engines arrived, and within a few years they were powering most airliners and military aircraft. But there is still a place for the piston engine in aviation, particularly in small airplanes.

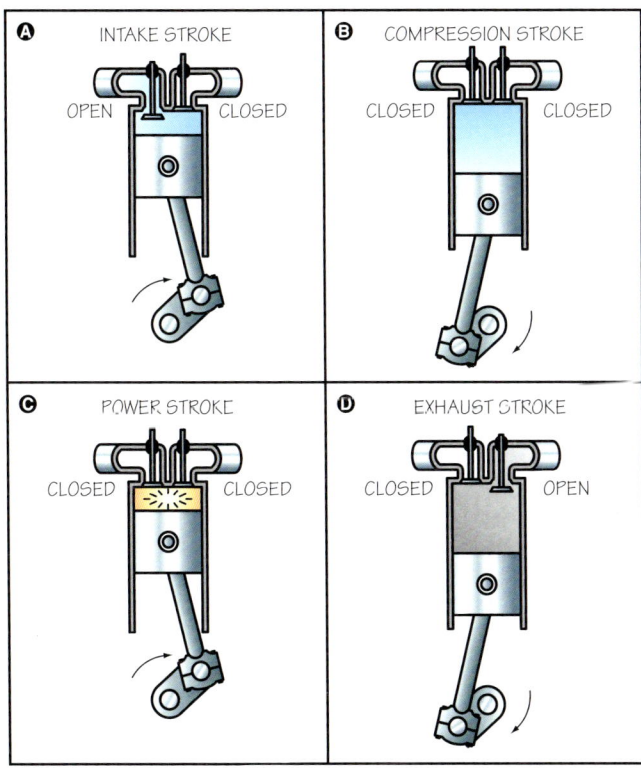

How a piston engine develops power

The Piston Engine

In a **piston engine** (or reciprocating internal-combustion engine), a piston compresses a mixture of gasoline and air. When an electric spark ignites this compressed mixture, the resulting gases expand very rapidly and force the piston to move away from the end of the cylinder in which it is enclosed. This motion is transferred to a connecting rod, which then transmits a rotary motion to a crankshaft. The rotating crankshaft turns the propeller and also forces the piston back to the top of the cylinder.

An exhaust valve at the top or head of the cylinder opens to let out the burned gases. It then closes, and another valve—the intake valve—opens to let in a fresh mixture of air and gasoline. These valves are operated by a system of gears and cams so that they open and close at the correct time.

Each complete movement of a piston in one direction is a **stroke.** The whole series of actions between the admission of gasoline and air and the exhaust of the burned gases is called a **cycle.** The main aircraft engines use a four-stroke cycle, so called because there are four strokes of the piston in each cycle. The crankshaft makes two complete revolutions for each four-stroke cycle.

Each cylinder in an engine fires once for every two revolutions of the crankshaft. The power strokes in the various cylinders are timed to keep the crankshaft turning smoothly. The propeller converts the power of the engine into the thrust that pulls an airplane rapidly through the air.

> Most light airplanes have four- or six-cylinder engines.

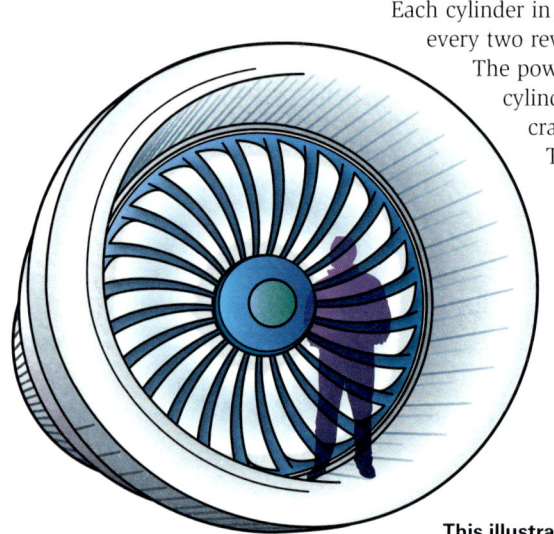

This illustration shows how large the engine of a commercial airplane can be.

Jet Propulsion

A simple experiment with a balloon demonstrates how jet engines work. Blow up the balloon and pinch the neck to keep in the air. Then, let the balloon go and watch it shoot across the room. The air inside the balloon is pushing in all directions to get out, but it can escape only through the open neck. The air at the opposite end of the balloon can't escape quickly, and its push against the front is what makes the balloon shoot forward. When all the air has escaped through the neck, there is no more jet propulsion, and the balloon falls in a rubbery heap.

The scientific principle behind this jet action is Newton's third law of motion, which was discussed earlier in this chapter. In our experiment with the balloon, the force of the air rushing from the neck creates an opposite, equal force that pushes the balloon forward. Of course, in a real jet or rocket engine, the thrust has to continue for a lot longer if our plane or rocket is to fly more than a few feet, but the principle is the same.

Jet Engines

Jet engines and rockets both operate on Newton's law, creating a thrust that pushes the craft through the air or into outer space. The difference between them is that jet engines require air, while rockets do not; rockets carry their own oxygen.

In jets, a stream of air and burning gases flows through the engine and is ejected at a speed much greater than when it entered. Most jet engines use rotating compressors and turbines to compress the gases and operate the moving parts. When the gases are compressed by scooping air from the atmosphere without compressors or turbines, the engine is called a ramjet.

There are five kinds of jet engines: turbojet, turbofan, turboprop, ramjet, and scramjet.

Turbojet. In a turbojet engine, the air comes in through an inlet and is compressed by the rotating blades of a compressor. It is then heated to tremendous temperatures by being mixed with the fuel (kerosene) in a combustion chamber.

The expanding gases travel through a rotating turbine and out the tail, creating a giant thrust. The turbine turns the compressor, which is usually mounted on the same shaft. The turbine does not contribute to the engine's thrust.

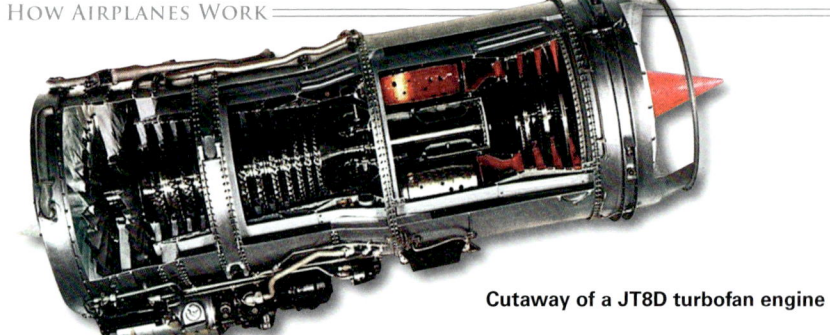

Cutaway of a JT8D turbofan engine

Turbofan. Turbofan engines are similar to turbojets except that they have larger rotating blades enclosed in a metal cowling. And in the turbofan, a separate portion of the compressor provides an air stream that bypasses the combustion chamber. The turbofan engine has greater thrust for its weight than the turbojet.

Turboprop. This is a variation of the turbojet. Some of the shaft power of the turbojet engine is used to turn a propeller, just like those on piston-driven aircraft. Turboprops get most of their thrust from the propellers, but they also are propelled by the exhaust gases. This added thrust is especially important at high altitudes and high speeds.

The turboprop engine is particularly useful in commercial aviation because the propeller gives a large thrust for takeoff. It can also be reversed easily and quickly for landings at slower speeds than regular jet aircraft.

Ramjet. The ramjet engine is the simplest power plant ever devised. The compressor and turbine are eliminated altogether, and it has no moving parts. The air inlet on a ramjet is designed so that the plane's forward speed alone compresses the air. It goes into a combustion chamber where it is mixed with fuel. The expanding gases go directly out the tail, creating a tremendous thrust.

Ramjets have one shortcoming: They cannot power the aircraft by themselves. Usually, a rocket or some other device sends the ramjet on its way. Ramjets can be designed to operate from slightly over Mach 1 (the speed of sound, about 761 mph at sea level) to Mach 4 or 5.

Scramjet. Scramjet stands for "supersonic combustion ramjet." This engine operates by introducing a mixture of fuel and air to fly at supersonic speeds (in excess of Mach 5). By capturing air into specially designed engine inlets, then compressing the air and combining it with the fuel in a supersonic combustor, the scramjet is able to achieve supersonic speeds. Today, there are many test vehicles demonstrating hypersonic flight—at speeds higher than Mach 5—using the scramjet engine.

Rockets

Rockets were used as fireworks and weapons of war by the Chinese at least a thousand years ago, but modern development of rockets did not begin until the first half of the 20th century.

The rocket engine is used in space flights because it carries its own oxygen, but rockets can also be used in the atmosphere. In fact, the first plane to fly faster than the speed of sound—the X-1—was rocket-powered. Today, rockets are important mainly as propulsion for spacecraft reaching into space.

If you want to learn more about rockets and humankind's future in outer space, considering earning the Space Exploration merit badge.

= All About Instrumentation

All About Instrumentation

The first aviators flew by the seat of their pants, relying on their senses to tell them how high they were and whether they were climbing or descending. This style of flying was pretty easy because early airplanes had open cockpits and flew both low and slow. However, as planes got faster and pilots started flying at night, they found they could not always trust their senses. So airplane builders began adding instruments to indicate the plane's altitude, its heading, and whether it was climbing or descending, turning or flying straight, in level flight or banking.

Modern airplanes have instruments and radios to tell the pilot everything about the aircraft's position and condition. With them, the pilot hardly has to look out the windshield to fly the plane. That is not to say that pilots do not look outside, however. In clear skies, they rely on both their instruments and their eyes to determine the plane's position. When flying in clouds, however, they must rely solely on instruments.

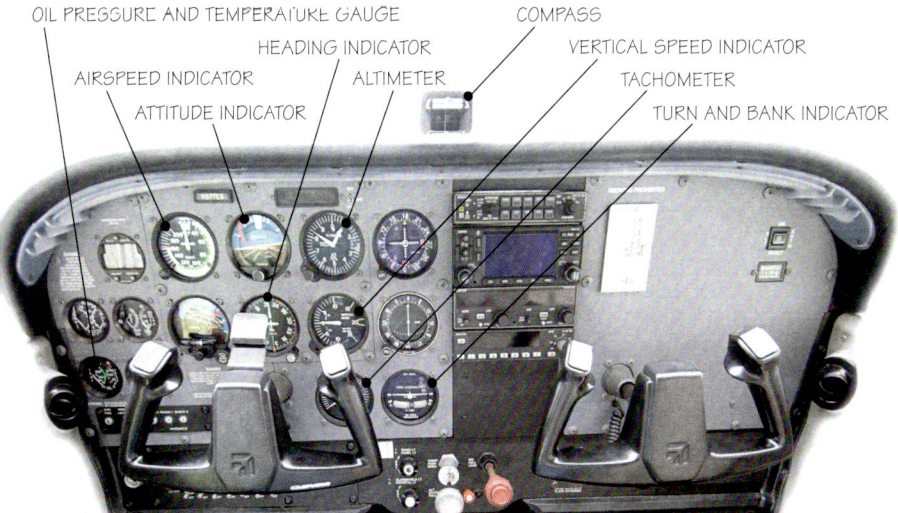

The seven flight instruments mentioned in requirement 2f—the attitude indicator, heading indicator, altimeter, airspeed indicator, turn and bank indicator, vertical speed indicator, and compass—tell the pilot about the plane's altitude, speed, direction, attitude, and rate of climb or descent. The navigation/communication radios (navcoms), discussed more in the next chapter, permit the pilot to guide the aircraft directly and safely to its destination in both clear and cloudy conditions. The last three instruments—tachometer, oil pressure gauge, and oil temperature gauge—tell a pilot how the engine is operating.

Let's look at the flight instruments first.

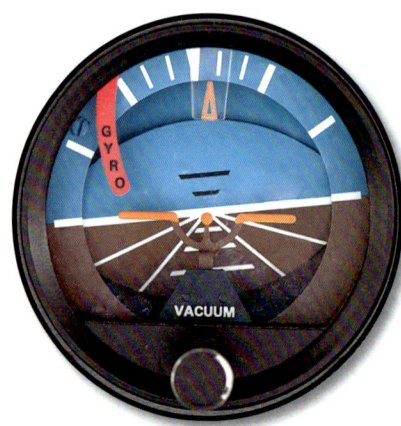

Attitude Indicator

The attitude indicator or "artificial horizon" lets the pilot get an immediate picture of the airplane's **attitude,** which is its position relative to Earth's horizon.

Attached to a gyroscope is a face with a contrasting horizon line on it. This line represents Earth's actual horizon. A miniature airplane on the housing moves with respect to this artificial horizon, just like the real plane moves with respect to the real horizon. The attitude indicator shows both bank (roll) attitude, which is the relationship between the wings and the horizon, and pitch attitude, which is the relationship between the nose and the horizon.

Heading Indicator

The heading indicator can take different forms but is basically a gyroscope that shows the plane's heading. The simplest heading indicators have to be set to match the magnetic compass. Others have their own compass, to which they are "slaved" (connected to and directed by the instrument), allowing the setting to be maintained automatically.

= All About Instrumentation

One reason for having a heading indicator in addition to a magnetic compass is that the compass does not read accurately when not in straight and level flight, whereas the gyroscope maintains an accurate and steady reading.

Altimeter

The altimeter tells the pilot how high the aircraft is flying. To steer clear of mountains, buildings, and such obstructions as television towers, the pilot must know the altitude at all times. Charts and air traffic rules indicate the minimum heights pilots must maintain. There are also specific altitudes to fly based on the direction of flight, which reduces the risk of collisions when pilots are flying by visual flight rules in good weather. During times of reduced visibility, pilots are assigned altitudes by air traffic control.

The altimeter is simply a barometer that measures the air pressure and converts that measurement into altitude. Altitude references are generally above sea level, so the pilot has to know the height above sea level of the terrain or obstructions to be sure the plane is at a safe altitude. The altimeter has a knob for adjusting the instrument to take into account changes in barometric pressure at different points of the flight as reported by weather stations.

As discussed earlier, atmospheric pressure gets lower as altitude increases.

AVIATION 35

All About Instrumentation

Airspeed Indicator

The airspeed indicator is the airplane equivalent of a car's speedometer, telling the pilot how fast the plane is traveling through the air. Like the altimeter, the airspeed indicator works by measuring air pressure, but the airspeed indicator measures the plane's impact on the air (ram air pressure). In other words, the airspeed indicator registers the velocity of air molecules striking a sensor as the airplane moves through the air. This is translated into speed in **knots,** or nautical miles per hour, the standard unit of velocity used in aviation.

Compass

As a Scout, you are familiar with the magnetic compass and probably have used one many times. The compass used in aircraft is not much different, although flight poses special problems in reading a compass. You may know that the magnetic pole is not at the North Pole, or the exact top of Earth. Instead, it is around 800 miles away, which leads to variations in determining true headings. Second, Earth is not uniformly magnetized; in some areas, the compass may vary many degrees from magnetic north. Finally, the metal and electrical equipment within an aircraft can throw off the compass. A pilot must consider variation and deviation, as well as wind, when determining what compass reading will get the airplane to its destination.

Turn and Bank Indicator

The turn and bank indicator is two instruments in one. It tells the pilot when the plane is turning and how well the turn is being executed—whether there is too much or too little bank for the rate of turn. The pilot may also check for balance and coordination in straight and level flight. The turn needle always deflects in the direction of the turn and indicates the rate at which the aircraft is turning about its vertical (yaw) axis.

Most modern airplanes have a variation of this instrument called a turn coordinator. It looks a little like an attitude indicator but gives information only about turn, not about pitch attitude. The ball part of the turn indicator is simply an agate or steel ball that moves freely inside a curved, sealed glass tube filled with liquid. The lowest point of the glass tube is in the middle of the instrument. In straight and level flight, gravity keeps the ball there, centered between two lines.

In a turn, if the aircraft is neither slipping nor skidding, the ball will be kept centered by centrifugal force. If the aircraft were in a slip (the tail sagging into the turn), the ball would fall to the low side of the instrument. In a skid (the tail swinging wide outside the turn), the ball would be to the high side.

Vertical Speed Indicator

In addition to knowing the airspeed, the pilot must know how rapidly the aircraft is climbing or descending. The vertical speed indicator, or VSI, registers how fast the barometric pressure is changing and converts this information to a speed measured in hundreds of feet per minute. This instrument is important because it is difficult to judge rates of climb or descent using only our human senses.

In addition to the flight instruments, the tachometer, the oil pressure gauge, and the temperature gauges also tell the pilot how the plane's engine is performing.

All About Instrumentation

Tachometer

You may have seen a tachometer on the dashboard of a car. Its purpose is to tell the driver exactly how fast the engine is running. In an airplane with a piston engine and a fixed-pitch propeller, the tachometer has two main purposes: to show whether the propeller is turning at the recommended speed for a particular maneuver and to indicate whether the engine is operating normally. For example, the airplane's designer might have determined that the best cruising speed for the engine is 2,300 revolutions per minute (rpm), so the pilot would set the throttle accordingly while cruising. The designer would also have recommended certain rpm settings for climbing and descending.

The tachometer also tells the pilot something about the engine's condition. Suppose a pilot is preparing for flight and finds that, with the throttle open all the way, the tachometer reads only 1,800 rpm when it should read 2,400. That is a good indication something is wrong with the engine.

Oil Pressure Gauge

An airplane's oil pressure gauge does the same thing as the oil pressure gauge in a car. It shows the pilot the pressure of the oil in the engine, which reveals a great deal about the health of the engine. Dropping oil pressure is a sure sign of engine trouble.

OIL PRESSURE

Temperature Gauges

TEMPERATURE

The temperature gauges are another indicator of the engine's health. They measure the temperature of oil and the cylinder heads and show whether the engine is running well, too warm, or too cold. The instruments are generally marked with a green area and a red line. If the needle is "in the green," that is good. If it passes the red line, there is a problem because that line marks the maximum allowable operating temperature.

Communication

Virtually all airplanes are equipped with radios, so a pilot is never out of range of advice and help. Good radio communication is vital to safe flying. The Federal Aviation Administration maintains control towers at airports and flight service stations throughout the country to provide dependable communications for pilots. These communications can take the form of clearances to take off, to land, or to fly under instrument flight rules (where air traffic controllers keep all aircraft separate).

Pilots can also get the latest weather information while aloft. To do this efficiently, a standard method of communicating has evolved. For example, if you were in the control tower at Redbird Airport, you might hear this exchange.

Pilot: Redbird tower, this is Cessna three-niner-two-one-bravo, 10 miles west, inbound for landing. Over.

Tower: Cessna three-niner-two-one-bravo, Redbird tower. Report a left base, runway three-six. Wind three-four-zero degrees at one-five. Altimeter two-niner-niner-five.

Pilot: Left base, runway three-six. Cessna three-niner-two-one-bravo.

Here is what that exchange would look like in plain English.

Pilot: Redbird tower, this is Cessna 3921B. I'm 10 miles west of the airport and am coming in for a landing.

Tower: Cessna 3921B, this is Redbird tower. Approach from the left base section of the designated initial approach area and land on runway 36. The wind is coming from 340 degrees at 15 knots. The barometric pressure is 29.95.

Pilot: OK. I'll do what you instructed and land on runway 36.

COMMUNICATION

When you compare these two versions, you will notice that the first version is about half as long as the second. It also includes some unfamiliar words and pronunciations.

Pilots use shorthand phrases that both they and the tower understand. They also use the International Civil Aviation Organization's phonetic alphabet, which is shown in the chart on the next page. This alphabet helps reduce misunderstandings, because difficult words can be spelled out with it or pronounced in a special way that is easy to understand.

Numbers are straightforward, but there are some special pronunciations. For example, nine is pronounced "ny-ner" because it is easily confused with five. Also, altitudes above 10,000 feet are referred to by their digits (such as "one-three thousand" for 13,000). However, when at or above 18,000 feet, altitudes become flight levels. For example, 22,000 feet is flight level 220 (pronounced "two-two-zero").

At or above flight level 180, all airplanes must fly under positive (mandatory) air traffic control and are separated vertically by at least 1,000 feet. Above flight level 290, they are separated by 2,000 feet because altimetry does not work as well at higher altitudes.

COMMUNICATION

ICAO Phonetic Alphabet

A	Alpha	J	Juliet	R	Romeo
B	Bravo	K	Kilo (Key-loh)	S	Sierra
C	Charlie	L	Lima (Lee-mah)	T	Tango
D	Delta	M	Mike	U	Uniform
E	Echo	N	November	V	Victor
F	Foxtrot	O	Oscar	W	Whiskey
G	Golf	P	Papa	X	X-ray
H	Hotel	Q	Quebec	Y	Yankee
I	India		(Keh-beck)	Z	Zulu

Common Aviation Terms

There are a few words and phrases that are not part of the phonetic alphabet that also help in radio communication. Here are some of the common ones and what they mean.

Acknowledge—Let me know that you have received and understood this message.

Affirmative—Yes.

Negative—No.

Correction—An error has been made in this transmission. The correct version is . . .

Over—My transmission is ended, and I expect a response from you.

Out—This conversation is ended, and I don't expect a response.

Roger—I have received all of your last transmission.

Read back—Repeat all of this message back to me exactly as received after I have said, "Over."

Aviation jargon might seem strange at first, but few words are actually used. English is the official language of aviation the world over, and it has been said that pilots who speak other languages have to learn only about 100 English words to operate in any air traffic control system.

Air Navigation

Air navigation is simply the way a pilot gets an airplane from where it is to its destination with a minimum of difficulty. Using charts, instruments, and preparation, the pilot is able to map a course so that he knows at all times where the plane is, what direction it should go, and how long the trip will take.

There are four standard methods of navigation.

- **Pilotage**—by reference to charts and visible landmarks
- **Dead (deduced) reckoning**—by figuring direction and distance from a known position
- **Radio navigation**—by use of ground-based radio aids
- **Satellite navigation**—by use of the global positioning system

Aeronautical Charts

A chart is what most people would call a map. There are four types of charts of greatest interest to private pilots.

Local aeronautical chart. The scale is 4 statute miles per inch. (A statute mile is 5,280 feet.)

Sectional aeronautical charts. Scale is 8 statute miles per inch. (This chart also includes scales for nautical miles and kilometers.)

World aeronautical chart. Scale is 16 statute miles per inch.

Aeronautical planning chart. Scale is 80 statute miles per inch.

Sectional Charts

The sectional aeronautical charts are most widely used by private pilots. There are 37 such charts for the 48 contiguous states; additional sectionals cover Alaska and Hawaii. For requirement 2c, you must obtain a chart and learn how to read it.

You should be able to find the sectional aeronautical chart for your area at your local airport. You can also obtain copies by writing to the National Oceanic and Atmospheric Administration. See the resources section at the end of this pamphlet.

Reading an aeronautical chart is almost like reading a detailed highway map. In fact, they look like road maps, except that aeronautical charts have special symbols to show such things as airports, radio navigation aids, latitude and longitude, and magnetic variation. Each chart includes a detailed explanation of the symbols used; your counselor can also help you understand what you see.

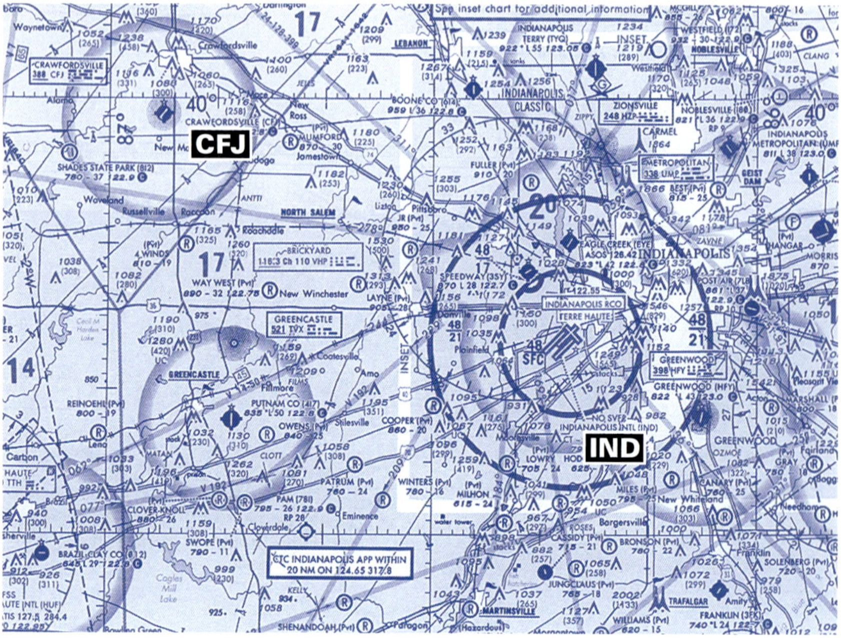

AIR NAVIGATION

The map here shows part of the St. Louis sectional aeronautical chart. We will use this part of the chart in explaining how a pilot would map a course for a short flight between two cities in Indiana.

Measuring Direction

Before you can plot a course, you should review some basic facts of geography. As you know, the equator is an imaginary line running around the middle of Earth exactly halfway between the North and South Poles; its value is zero degrees. Other imaginary east-west lines that circle Earth are called **lines of latitude,** or **parallels.** This country's 48 contiguous states lie roughly between 25 degrees and 49 degrees north latitude (that is, north of the equator).

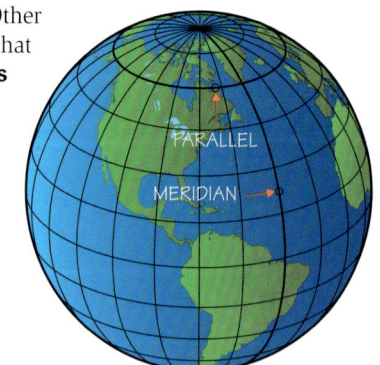

46 AVIATION

> You can locate any point on Earth by reference to its latitude and longitude. For example, Washington, D.C., lies at approximately 39 degrees north latitude, 77 degrees west longitude.

Lines of longitude, or **meridians,** are imaginary lines that run from the North Pole to the South Pole at right angles to the equator. The meridian passing through Greenwich, England, is called the prime meridian and has a value of zero degrees. Other meridians have values in degrees of longitude east and west up to 180 degrees. North America lies between about 67 and 125 degrees west longitude.

Approximate measures are not very helpful, however, when you consider that the distance between two degrees of latitude is roughly 69 miles. So each degree is subdivided into 60 minutes, which lets us pinpoint a location further. On charts, degrees are shown by this symbol (°) and minutes by this (').

On the part of the sectional aeronautical chart shown on the facing page, Indianapolis International Airport is at 39 degrees 43 minutes north latitude, 86 degrees 17 minutes west longitude. You can't see the latitude in the illustration, however, because it does not include the edge of the chart where the latitude appears.

The True Course

Let's suppose that we are at Indianapolis International Airport (IND on the chart), and we want to fly directly to Crawfordsville Airport (CFJ), which is 33 nautical miles to the northwest. IND, as we said, is at 39 degrees 43 minutes north latitude, 86 degrees 17 minutes west longitude, while CFJ is at 39 degrees 58 minutes north latitude and 86 degrees 55 minutes west longitude.

To determine the direction to Crawfordsville Airport, we draw a line on the chart from IND to CFJ. At a meridian line near the first third of our flight, we measure the angle between our course and the meridian, using a plotter. This angle is 298 degrees, and we call it the **true course** because it measures a direction with reference to true north. True course for our return trip would be 118 degrees, or the reciprocal of 298 degrees (298 degrees minus 180 degrees).

AIR NAVIGATION

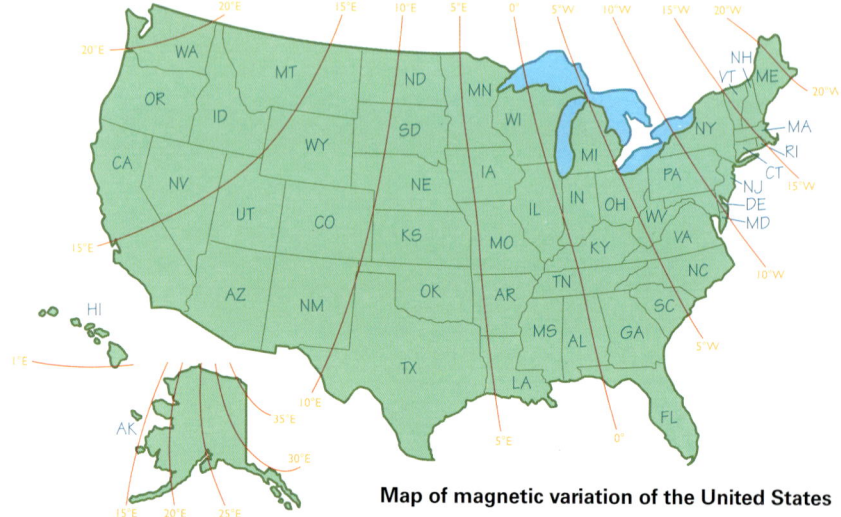

Map of magnetic variation of the United States

Magnetic Variation

You might think that all we have to do is take off from Indianapolis International and steer a course of 298 degrees to Crawfordsville Airport. But it is not quite that easy, because we have other factors to consider.

The first is magnetic variation. As we discussed in the chapter on instruments (and as you may have learned from work on the Orienteering merit badge), compass needles point to magnetic north, not to the geographical North Pole, which is about 800 miles away. In the United States, true north and magnetic north coincide on a line running roughly through Wisconsin, Illinois, Kentucky, Tennessee, and Alabama. This is called the **agonic line** (see the chart of the United States showing magnetic variation across the country). Unless you are on the agonic line, you must compensate for the difference between true and magnetic north, which can be more than 20 degrees.

Because Earth is not uniformly magnetized, in some places, a compass needle will vary many degrees from magnetic north because of variation in Earth's magnetism. The U.S. Coast and Geodetic Survey has carefully measured this variation at all parts of the country. On aeronautical charts, the amount and direction of variation is shown by broken red lines called **isogonic lines.**

Near Indianapolis, the magnetic variation is approximately 2 degrees west, which means we need to add 2 degrees to our course. If the variation were to the east—as it would be in most of Illinois—we would subtract that value. (An easy way to remember whether to add or subtract is the rhyme "East is least, west is best.")

So, allowing for magnetic variation, our course from IND to CFJ would be 300 degrees. On the return trip, we would add 2 degrees to our true course of 118 degrees for a magnetic course of 120 degrees.

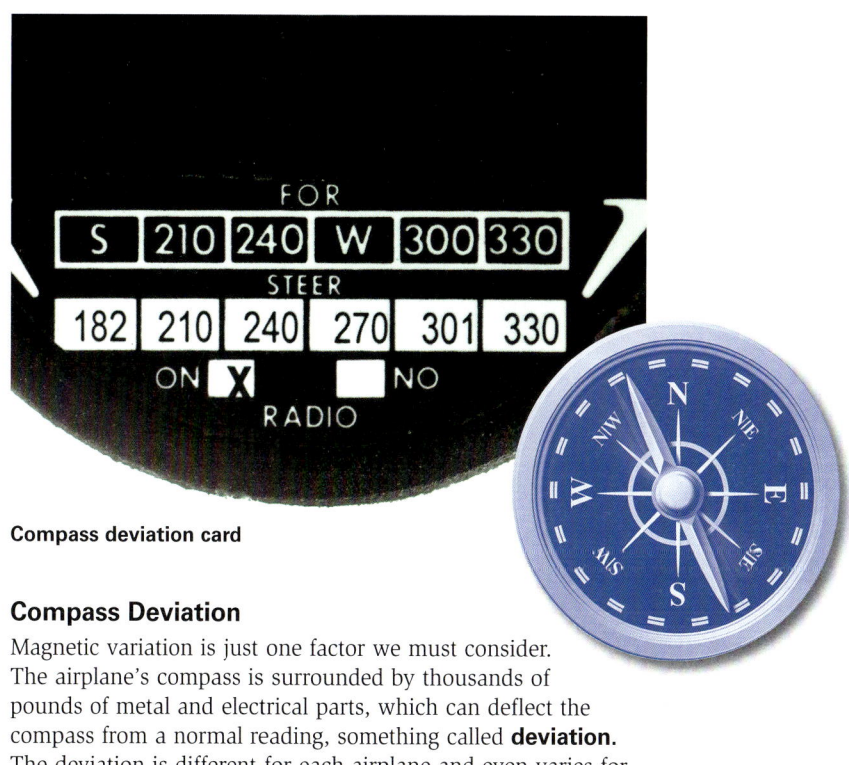

Compass deviation card

Compass Deviation

Magnetic variation is just one factor we must consider. The airplane's compass is surrounded by thousands of pounds of metal and electrical parts, which can deflect the compass from a normal reading, something called **deviation.** The deviation is different for each airplane and even varies for the same airplane on different headings.

Air Navigation

> Technicians can calculate closely just how much the deviation will be for any airplane. In every aircraft, you will find a deviation card mounted near the compass that tells us how to correct the magnetic course for deviation.

Remember, the magnetic heading from Indianapolis International (IND) to Crawfordsville Airport (CFJ) is 300 degrees. Our deviation card tells us that for a magnetic course of 300 degrees, we should steer a course of 303 degrees. In other words the compass deviation in our airplane at this course is 3 degrees higher than the magnetic course.

We can now proceed to steer at 303 degrees, confident that we will soon see the Crawfordsville Airport. And we would—provided that there was no wind.

Wind Drift

For requirement 2f, you learned that your airspeed indicator shows how fast you are passing through the air, not over the ground. If you are flying into a 25-knot headwind, you are actually traveling over the ground at 25 knots less than your airspeed. On the other hand, if you have a 25-knot tailwind, you are traveling over the ground 25 knots faster than your airspeed.

However, if the air mass is moving across your path, then your plane will drift with the wind. As a pilot, you must know how to correct your course for this drift, which you do by using your own speed and the speed and direction of the wind to make a **wind triangle**. When the course is corrected for wind, it is called a **heading**. That is what you fly on the compass to reach your destination.

Now let's draw a wind triangle using our calculations for a planned flight from IND to CFJ.

AVIATION

AIR NAVIGATION

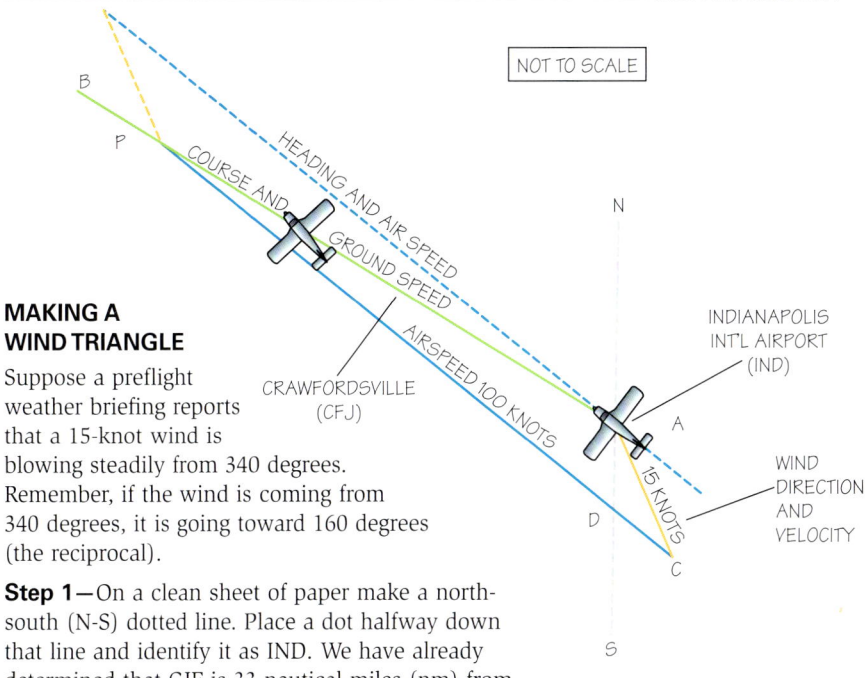

MAKING A WIND TRIANGLE

Suppose a preflight weather briefing reports that a 15-knot wind is blowing steadily from 340 degrees. Remember, if the wind is coming from 340 degrees, it is going toward 160 degrees (the reciprocal).

Step 1—On a clean sheet of paper make a north-south (N-S) dotted line. Place a dot halfway down that line and identify it as IND. We have already determined that CJF is 33 nautical miles (nm) from IND and that our true course is 298 degrees.

Step 2—Using a protractor, measure our 33 nm true course of 298 degrees based on our N-S line (A-B). Because our plane's airspeed is 100 knots (nautical miles per hour), make A-B 100 units long. Place a dot 33 units from the N-S line and identify it as CJF.

Step 3—With the protractor centered on the dot representing IND, draw a line at 160 degrees clockwise from north, indicating the effects of the wind from 340 degrees to 160 degrees. Since it is a 15-knot wind, show its velocity by making that line 15 units long on our ruler; place an arrowhead (>) at the end of that line and identify it as C.

Step 4—The plane's cruising speed is 100 knots, so measure 100 units on the ruler to represent the airspeed, making a dot on the ruler at that point. Place the ruler so that the end is on the arrowhead and the 100-mile dot on the ruler crosses the true course line (label that point P). Draw the line and label it for our 100-knot airspeed. At the end of an hour, our position should be at point P.

AVIATION   51

AIR NAVIGATION

Because weather constantly changes, we must check our calculations during the flight by radio or reference points on the ground to determine if the wind has changed in its direction and/or speed.

Step 5 — We can find our ground speed by measuring the number of units on the ruler from Indianapolis International Airport to point P, which is 88 knots.

Step 6 — What we really want to know, however, is what our heading should be to offset wind drift on a flight from IND to CFJ. We can find that from our wind triangle by placing the reference line of the protractor along the north-south line with its center point at the intersection of the airspeed line (D). We find that the airspeed line extends out at 304 degrees, our true heading. To calculate the compass heading, which is what we will actually fly, we must also consider the effects of magnetic variation and compass deviation.

Step 7 — After we obtain this true heading, we can apply the corrections for magnetic variation and for compass deviation to obtain a compass heading, as we discussed earlier. This will enable us to fly directly to CFJ. We have already found that magnetic variation in the Indianapolis area is 2 degrees west, so we add 2 degrees to our true heading and find a magnetic heading of 306 degrees.

Step 8 — The deviation card near our compass tells us to add 3 degrees to a magnetic reading of 300 degrees, so we add that to our magnetic heading and get 309 degrees. That is our compass heading from IND. It should take us directly over CFJ.

AIR NAVIGATION

The Wheel Deal

Drawing a wind triangle is a pretty accurate way to figure out your heading, but it can be awkward. Not surprisingly, pilots have tools to make the job easier.

One of these tools is the E6B flight computer. Sometimes called a "whiz wheel," the E6B has been around for generations. One side is a circular slide rule marked for performing flight calculations. The other side—the "wind" side—has a transparent sheet of plastic over a grid. In effect, you draw your wind triangle on the clear plastic of the computer instead of on a piece of paper. Grid lines and other markings on the computer let you simply read off angles and speeds without having to measure them with a protractor and ruler.

Learning to use an E6B takes practice but is a mark of a knowledgeable aviator. Electronic flight computers similar to handheld calculators are available and work quite well, but the well-prepared pilot knows how to use the old E6B, too. This instrument is vital to a pilot's knowledge because it gives a mental image of the relationships of performance and winds—something electronic units cannot do.

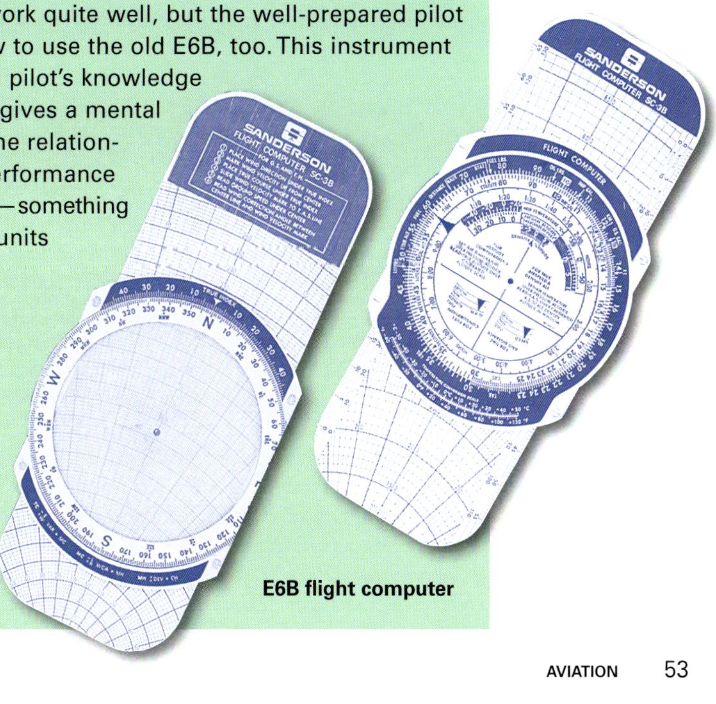

E6B flight computer

AVIATION 53

Radio Navigation: VOR Highways

The Federal Aviation Administration oversees radio aids to navigation through its Federal Airways System. The chief ground-based aid is called very high frequency, omnidirectional radio range ("VHF omni range" or "VOR"). From VOR stations throughout the country, VHF radio signals are transmitted in all directions, or *omnidirectional*. Any aircraft equipped with a VOR receiver can receive these signals if it is within range. Depending on the VOR station's power and the aircraft's altitude, range can be more than 100 miles. Information on distance may also be available if both the ground station and aircraft are equipped with distance measuring equipment.

Assume that you are a pilot and want to use VOR as a simple navigation guide. Look at the aeronautical chart covering your destination, find the location and radio frequency of a nearby VOR station, and tune to that station. Be sure you have the correct station—listen to the identification, which will be sent either in Morse code or by voice.

Turn the omni bearing selector (a knob at the edge of the VOR receiver's face) until the vertical needle centers at the bottom of the round dial. Glance at the "to/from" indicator; be sure the bearing is "to" the VOR. Now, just fly a course that keeps the vertical needle centered. If it moves right, fly right to correct your course. If it moves left, fly left. Keep the needle centered and you will fly directly over the VOR station. At that point, the "to/from" needle will flicker for a few seconds and then change to a "from" reading. To continue past the station in the same direction, keep the needle centered on "from."

When using VOR, remember that the needle indication shows your relationship to an imaginary line on the ground. When a correction is needed, try about a 15-degree correction toward the deflection of the needle. Then, when the needle returns to the center, subtract 10 degrees of that correction and maintain that heading to see if it will keep the needle centered.

With VOR, the pilot need not be concerned with magnetic variation, compass deviation, or wind drift. Keep the needle centered; radio does the rest.

GPS: Steering by Satellite

You may have used a global positioning system, or GPS, receiver on a Scout hike. GPS uses a series of geostationary satellites orbiting at an altitude of about 11,000 to precisely indicate your position in terms of latitude and longitude.

GPS is quickly becoming the primary means of navigation. It offers similar guidance to VOR/DME but can also accurately determine altitude. Because the GPS receiver knows your present position and your past position, it gives a very accurate ground speed readout as well as a readout of your track over the ground.

Eventually, the GPS system will replace the instrument landing system, which for years has been used for guidance to runways when the ceiling and visibility are very low. It probably will also replace the entire VOR system.

Entering the Cockpit

Now that you understand how airplanes work, you may be eager to get in the cockpit and start flying. Of course, not just anyone is allowed to fly an airplane; you must earn a pilot's license first, just as you must earn a driver's license before you can drive a car. The good news is that earning a pilot's certificate is not as hard as you might think.

The U.S. government's Code of Federal Regulations (Title 14) details five levels of pilot certificates, ranging from student pilot to air transport pilot (the certificate held by airline captains). To help you complete requirement 1e, we will briefly examine two of the certificates—recreational pilot and private pilot—along with the instrument pilot rating.

All of these certificates require an individual to write the English language, pass written and oral examinations, and be in good health. (Keeping yourself physically strong and mentally awake can come in handy here.) Lastly, you must be at least 16 years old to fly solo and at least 17 to carry passengers.

Recreational Pilot Certificate

Of all the certificates and ratings, the recreational pilot certificate is the easiest to obtain. It requires 30 hours of flight instruction (three hours of which must be solo flight time). The main privileges afforded the recreational pilot are that the pilot may carry one passenger (in a single-engine land-based airplane with up to four seats and 180 horsepower) and may share expenses with that passenger. The recreational pilot may fly in good visibility at altitudes of up to 10,000 feet above sea level or up to 2,000 feet above ground level, whichever is greater, and may travel as far as 50 nautical miles from the home airport. The recreational pilot may earn additional privileges.

Private Pilot Certificate

At one time considered to be the primary pilot's license, the private pilot certificate is still greatly desired. Its requirements are more demanding than those for the recreational pilot certificate, and its privileges are greater. This certificate requires a minimum of 40 hours of flight instruction (of which 10 hours must be solo), including cross-country experience (trips of more than 50 nautical miles), night flight, instrument flight, and radio communications. A private pilot certificate permits the pilot to carry more than one passenger, to fly higher than 10,000 feet above sea level, to go farther than 50 nautical miles, to fly at night and into airspace controlled by air traffic control, and to share expenses with passengers.

> The recreational pilot certificate has some important restrictions. Most importantly, the pilot can fly only during daylight hours and in airspace where he does not need to communicate with air traffic control.

Instrument Rating

With additional instruction, the private pilot can add **ratings** to a certificate—upgrades that allow privileges like flying multiengine aircraft or seaplanes. One rating many private pilots work to earn as quickly as possible is the instrument rating, which allows them to fly at times when instrument flight rules are in effect. Without the instrument rating, a pilot can fly only when visibility is good and visual flight rules are in effect. Of course, it is sometimes necessary to fly when clouds obscure the sky. The instrument rating permits a pilot to fly under these conditions.

Taking Flight

You have learned the basic principles of flight and something about how to fly an airplane. You also know a little about what it takes to become a pilot. Perhaps you have never actually flown. Now is a good time to try if that is possible. Learning by doing is the Scouting way, and you cannot experience the thrills of flight while sitting on the ground.

You may be lucky enough to fly with your parent on a business trip or perhaps take a flying vacation with your family. Or, you may be able to take an orientation flight at an airport near your home—with your parent's permission, of course. If you are able to take an actual flight in an airliner or a small airplane, be sure to note the date, time, place, and how long the flight lasted— much as pilots log every flight they make. Be sure to make notes of your experience so you can tell your counselor about it.

Planning an Airplane Trip

Requirement 2e asks you to plan an imaginary airplane trip to at least three different locations. Here is your chance to dream a little. You could see the twinkling night lights of New York, London, or Paris; soar over the Alps or the Rockies; or look down on the Mediterranean Sea. You might stop for a day of sightseeing in Chicago or spend a couple of days in Tokyo. Perhaps you would rather spend a long weekend in Tahiti.

When you have decided where you would like to go, the next thing to do is plan your route and schedule. You do this by using airline timetables, which you can get at a travel agency, any commercial airport, the offices of airlines in big cities, on the Internet (with your parent's permission), and at many large hotels. If you live far from a big city, your best bet for help in planning an air cruise might be a local travel agency.

Using a computer, a travel agent can help you plan your route and schedule, check to see whether seats are available on the flights you want to take, and "finalize" your reservation. You can also do the same thing yourself via the Internet, either on an airline's Web site or on sites that handle reservations for multiple airlines.

You can reach almost any part of the world on regularly scheduled flights. All large cities and many smaller ones are linked by the major airlines, with the large cities often referred to as **hubs**. Increasingly, other cities and larger towns are being linked to major hub airports by smaller airlines that use turboprop or smaller jet airplanes called **commuters.** Their routes are called **spokes,** thus the term **hub and spoke system.**

For certain flights between major cities, some airlines do not require reservations. These shuttle runs guarantee anyone who wants to fly between these cities can do so just by coming to the airport on time. If more than one airplane is needed to handle the number of passengers who appear, the airline puts another plane into service.

Filing a Flight Plan

If you are a pilot, preparing for a trip is more complicated than just buying a ticket. You must plan every detail of the flight, from what time you will leave to how much fuel the journey will take.

Pilots flying under instrument flight rules are required by law to file an aviation flight plan—and can be fined up to $1,000 if they do not. But it is a smart idea to file a flight plan even when visual flight rules are in force.

Requirement 2e asks you to complete a flight plan for your imaginary airplane trip. You can use the sample form provided as a model.

FLIGHT PLAN

1. Type	2. Aircraft Identification	3. Aircraft Type / Special Equipment	4. True Airspeed	5. Departure Point	6. Departure Time (Z)		7. Cruising Altitude
VFR / IFR / DVFR			KTS		Proposed	Actual	

8. Route of Flight

9. Destination (name of airport and city)	10. Estimated Time En Route		11. Remarks
	Hours	Minutes	

12. Fuel On Board		13. Alternate Airport(s)	14. Pilot's Name, Address, and Telephone Number and Home Base	15. Number Aboard
Hours	Minutes			
			17. Destination Contact/Telephone (Optional)	

16. Color of Aircraft	CIVIL AIRCRAFT PILOTS. FAR 91 requires you file an IFR flight plan to operate under instrument flight rules in controlled airspace. Failure to file could result in a civil penalty not to exceed $1,000 for each violation (Section 901 of the Federal Aviation Act of 1958, as amended). Filing of a VFR flight plan is recommended as a good operating practice. See also Part 99 for requirements concerning DVFR flight plans.

CLOSE VFR FLIGHT PLAN WITH _____ FSS ON ARRIVAL

= TAKING FLIGHT

Preparing to Fly

A lot of preflight planning goes into every flight. The pilot must learn about the weather and the expected wind at the proposed cruising altitude. The pilot first determines whether everything is OK for the flight, and then conducts a preflight inspection of the airplane to make sure it is airworthy.

Follow along on a typical preflight inspection of a popular four-seat airplane. The pilot first visually checks the aircraft for general condition during a walk-around inspection. In cold weather, the pilot does this by removing even the smallest accumulations of frost, ice, or snow from the wings, tail, and control surfaces. The pilot then makes sure that the control surfaces contain no internal accumulations of ice or debris. If it is a night flight, the pilot checks the operation of all lights and makes sure a working flashlight (with extra batteries) is on board.

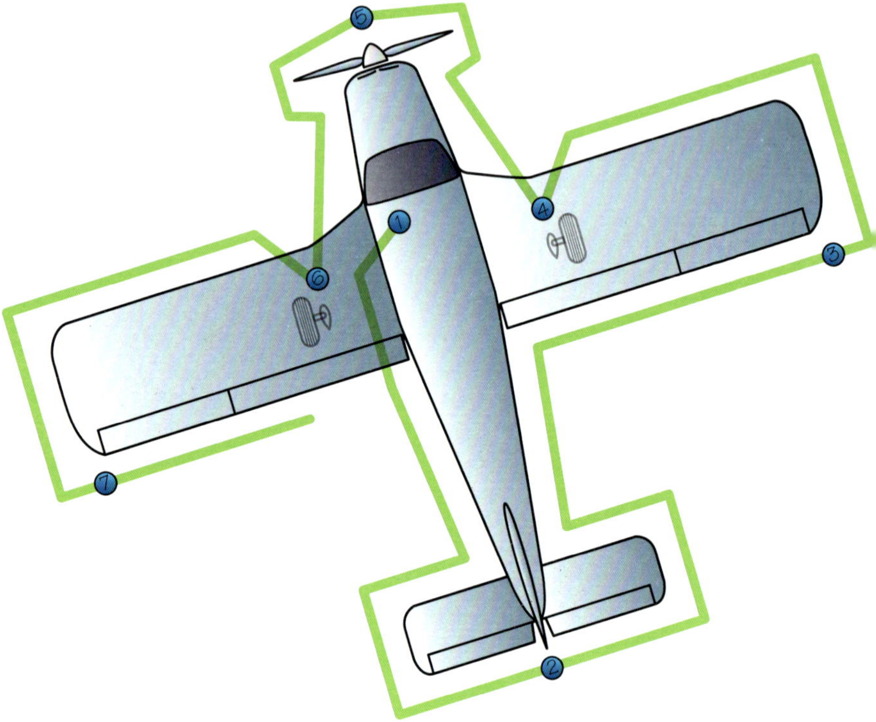

Here is a typical preflight checklist.

1. Cabin
   a. Remove the control wheel lock.
   b. Make sure the ignition switch is in the OFF position.
   c. Turn on the master switch and check fuel quantity indicators, then turn off the master switch.

   > Warning: When turning on the master switch, do not allow anyone to stand within the arc of the propeller, since a loose or broken wire or other malfunction could cause the propeller to rotate.

   d. Check the fuel selector valve handle on both tanks.
   e. Check baggage door for security.

2. Exterior, starting at the tail of the airplane
   a. Remove the rudder gust lock, if installed.
   b. Disconnect the tail tie-down.
   c. Check control surfaces for freedom of movement and security.

3. Right wing (trailing edge)
   a. Check the aileron for freedom of movement and security.

4. Right wing (leading edge)
   a. Disconnect the wing tie-down.
   b. Check the main wheel tire for proper inflation.
   c. Visually check the fuel quantity and that the fuel filler cap is secure.
   d. Before the first flight of the day and after each refueling, drain **all** fuel strainers, including the fuel line quick-drain valve. Drain at least a cupful of fuel (using a sampler cup), checking for water, sediment, and proper fuel grade. If contaminated, take further samples until clear, then gently rock the wings to move any additional contaminants to the sampling points. Take repeated samplings from **all** drain points until **all** contamination has been removed.

5. Nose
   a. Check the oil level. Do not operate a plane with the oil level below the recommended level. For extended flights, fill the oil to the maximum level.
   b. Check the propeller and spinner for nicks and security.
   c. Check the landing lights for condition and cleanliness.
   d. Check the carburetor air filter for restrictions by dust or other foreign matter.
   e. Check the nose wheel strut and tire for proper inflation.

6. Left wing (leading edge)
   a. Disconnect the tie-down rope.
   b. Inspect the flight instrument static source opening on the side of the fuselage for stoppage (left side only).

c. Check the main wheel tire for proper inflation.

d. Visually check the fuel quantity and that the fuel filler cap is secure.

e. Inspect the fuel strainer; repeat as with right wing.

f. Remove the pitot tube cover, if installed, and check the tube opening for stoppage. (The pitot tube measures air pressure and works with the air speed indicator.)

g. Check the fuel tank vent opening for stoppage.

h. Check the stall warning for stoppage.

7. Left wing (trailing edge)

a. Check the aileron for freedom of movement and security.

After the pilot completes a preflight check and boards the airplane, the checking continues. There is a checklist for every phase of flight, and the pilot uses this list to make sure everything is properly set before moving on to the next flying task.

Safety on the Ground and in the Air

If you are going to take a flight:

- Obey all instructions of the pilot and crew.
- Fasten your seat belt securely for takeoffs and landings and when instructed to do so. It is a good idea to keep your seat belt fastened whenever the airplane is moving—in the air or on the ground.

If you are walking around aircraft or hangars:

- Keep well away from propellers and helicopter tail rotors, even if they are not turning.
- Keep away from jet intakes and exhausts.
- Heed all warning signs.

As an airport visitor:

- Keep off the flying field.
- Stay out of operations rooms and the control tower unless you are being conducted by a guide.

- Remember that we live in an era of heightened security concerns. At major airport or radar facilities, expect to pass through tight checkpoints where you may need to empty your pockets.
- Never walk on an active tarmac; the noise may distract you from another danger you can't see.

Aviation Facilities

Every year, roughly 87,000 airplanes take off and land in the United States. At any given moment, about 5,000 planes are in the air. Yet, the number of aviation accidents (and near misses) is so small that every plane crash is a major news event.

How are so many aircraft able to operate so frequently yet remain safe and worry free? The answer lies in the air traffic control system, which guarantees trouble-free skies. This system includes airports and other facilities that may be unfamiliar to you.

> In the 1920s, the first attempts to control the airspace above the United States were made in establishing airmail routes. The organizations created back then were predecessors to the Federal Aviation Administration, which controls the skies today.

One of the earliest methods of navigation assistance for pilots was the lighting of fires at night. The first air traffic control used maps, chalkboards, and mental calculations to keep planes separated in the air. We have come a long way since then in providing accurate and reliable assistance for private, commercial, and military pilots.

> Requirements 4a and 4b encourage you to visit an airport and one of several types of FAA facilities. Before you go, it might be wise to learn a little about each.

AVIATION FACILITIES

The Airport

Visiting an airport should be fairly easy, because there is probably one not many miles from your home. Perhaps it is just a large, open, grassy field with a barnlike hangar for storing a few airplanes, or perhaps it is a giant complex near the center of a major city. In either case, a visit will give you a taste of aviation.

If possible, visit an airport that has an air traffic control tower. If you can visit the tower, you will be able to see the airways system at work as controllers guide traffic on the aerial highways.

Civil (commercial) airports are generally built by cities, often with aid from state and federal governments, to serve civilian aviation and sometimes military traffic as well. They are usually close to the city or town they serve. Their location may depend in part on where builders can find a reasonably level tract of land that is large enough to permit operations free of obstructions and prevailing smoke, fog, or strong winds.

70   AVIATION

Aviation Facilities

Airports are assigned identifiers consisting of three letters, numbers, or a combination of both. Although these identifiers sometimes seem random, they are usually drawn from the airport's name or city. Here are a few examples: Los Angeles International Airport is LAX, Hartsfield-Jackson Atlanta International Airport is ATL, and John F. Kennedy International Airport is JFK.

Airport Runways

At a large airport (and some small ones), the runways have hard surfaces of concrete or asphalt. Open areas are grass-covered to prevent erosion and to control dust. While light airplanes sometimes use the turf area for landings and takeoffs, airliners and other large planes need paved runways and taxi strips.

Airport runways are designed, whenever possible, to allow takeoffs and landings into the prevailing wind. All runways are numbered according to the compass heading indicated when lined up on the centerline of that runway, but in a sort of shorthand. So, for instance, a runway pointing toward 30 degrees would be runway 3.

Aviation Facilities

Runway numbers are pronounced one numeral at a time, so runway 21 is pronounced "two-one," not "twenty-one." Runway 30 is pronounced "three-zero," not "thirty."

If you landed on the same runway going in the opposite direction, the compass heading would be 210 degrees (the reciprocal of 30 degrees). Again, you would drop the last zero, referring to runway 21 instead of runway 210.

The runway in use (active) is determined by the wind direction; it is better to fly into the wind to reduce takeoff and landing roll. Some of the biggest airports have two or more parallel runways that can be used simultaneously. An airport with two runway 36s would designate them runway 36L (for left) and runway 36R (for right).

Other Airport Facilities

In addition to runways and a control tower, every large commercial airport has an administration building with passenger waiting rooms, restrooms, ticket counters, baggage areas, and offices for dispatchers, airline officials, and the airport management. The administration and airline terminal buildings may include restaurants, newsstands, rest and eating quarters for air crews or transient (visiting) private pilots, and other services. There will probably be taxi stands, bus stops, and rental car companies, allowing travelers to get from the airport to other destinations.

At large airports, each airline may have its own terminal facilities and a hangar or two for its airplanes. Fuel trucks and line crews are available. If the airport is served by an airline, there will be complete firefighting and rescue equipment. Other buildings will house equipment for maintaining the field.

AVIATION FACILITIES

The Control Tower

As its name suggests, a control tower is an elevated structure from which airport controllers operate. The people in the control tower have four primary responsibilities. They must

- Ensure aircraft are a safe distance apart and in the proper sequence when flying through the airport traffic area or in the traffic pattern.

- Manage arrivals and departures.

- Control ground movements of aircraft and ground vehicles.

- Provide clearances and local weather and airport information to pilots.

Terminal Radar Control Facility

More commonly known as approach/departure, the airport-based radar control facility provides separation for aircraft flying within 40 miles of the primary airport in the area. It may also be responsible for aircraft in that area up to 17,000 feet above the ground. The actual unit may or may not be located in the control tower or even on the airport site.

> When you visit a terminal radar control facility, expect to enter a darkened room lit only by the glow of radar screens. Controllers skillfully observe and maneuver the "blips" that appear on their screens; each blip represents an aircraft. If you have ever played a computer or video game, you have some idea of the intense nature of this responsibility. Of course, there is absolutely no room for error.

Air Route Traffic Control Center

Away from the airspace of a control tower, a pilot may receive assistance from an air route traffic control center, more commonly referred to as the "center." A controller's duties are chiefly oriented toward aircraft flying by instruments (instrument flight rules, or IFR) but, as the workload permits, a controller also assists those flying visually.

In addition to keeping aircraft a safe distance from one another, controllers monitor aircrafts' progress over their routes, advise pilots concerning hazardous weather, and sequence traffic into destination airports. Looking much like the approach/departure facility, air route traffic control centers may be responsible for much larger areas, some extending hundreds of miles from their building.

Flight Service Station

Providing weather and information services for all pilots is the primary responsibility of the flight service station. The FSS will also

- Accept and close flight plans.
- Conduct preflight weather briefings.
- Communicate with VFR pilots en route.
- Help pilots in distress.
- Provide weather information.
- Monitor air navigation radio aids.
- Publicize notices to airmen, which update pilots on changes to aeronautical facilities, services, or procedures or notify them of hazards.
- Work with search-and-rescue groups in locating missing aircraft.

Of all the FAA facilities mentioned in this section, the flight service station may be the most difficult to visit. The reason for this is the FAA's conversion to 61 automated flight service stations. However, you may be able to find one within a short drive of your hometown.

Flight Standards District Office

The Flight Standards District Office (FSDO, pronounced "fiz-doh") is a field office of the FAA serving the aviation industry and general public on matters relating to certifying and operating airline and general aviation aircraft. Each office serves an assigned geographical area and oversees operations for safety, certification of flight crews and aircraft, accident prevention, investigation, and enforcement.

Arranging a Visit

All of the facilities mentioned welcome visitors, and you can find their phone numbers in your phone book. Look up the airport's name or look for FAA facilities under "United States Government, Department of Transportation." Be sure to call ahead to make an appointment; don't expect to drop by unannounced. Airports, and especially control towers, are busy places, and the working personnel will not be able to drop everything to give you a guided tour without advance notice and permission.

Remember that you are a guest in a most important part of the air traffic control system. The people there are responsible for the lives and safety of thousands of individuals and billions of dollars in property.

Flying Without Leaving the Ground

Nothing quite equals the thrill of flying, but you can still enjoy aviation without leaving the ground. Computer simulators and model airplanes offer fun, inexpensive alternatives to taking an actual flight.

Flight Simulators

You may already have access to a flight simulator program for a personal computer. The equipment most likely includes a joystick or yoke control and may even have rudder pedals. If you choose option 2d, you will need to plot a course over an area contained in the program menu and obtain an aeronautical sectional chart for that region.

From the airplanes available in the menu, choose one that you want to fly. This exercise should not be viewed as entertainment, competition, or combat. The purpose of this requirement is to let you experience some sense of what it is like to pilot an airplane from takeoff to landing, to fly a heading that you calculated during the preflight stage, and to monitor your progress as you pass ground and radio checkpoints.

> Select a departure point and a destination point that will allow for a flight of no more than an hour. Arrange to be uninterrupted during your time at the controls, and be aware of what you are thinking and feeling as you perform the flight. When you complete the flight, write down your experiences and share your observations with your merit badge counselor.

Model Airplanes

A great way to learn about airplanes and the principles of flying is to build model planes. Model airplanes are not toys. They hold an important place in the history of flight and were actually the first heavier-than-air machines to fly. Engineers still use models today as they develop new airplane designs and test them in wind tunnels. You will find model building fun, and you will learn much about the science of flight.

There are two basic kinds of model airplanes—nonflying display models and flying models. Display models are miniature copies of full-size airplanes, precise in scale and exact in detail. You can buy plastic kits for these models at hobby and toy stores. Some beginner's kits include all the materials necessary, including glue, paint, and decals of the plane's markings. You will just need to furnish a sharp knife and a paintbrush.

Model Airplanes That Fly

The simplest kind of flying model is a glider. However, if you choose to make a flying model for requirement 3a, it must be a powered rise-off-ground model—it must have a landing gear and be able to start from the ground or floor and fly. The power source might be liquid fuel or a battery pack. You may use a kit or design your own.

Some flying models are rather simple. Many are made almost entirely out of balsa wood. In some of the more realistic models, a balsa framework for the fuselage (body), wings, and tail surfaces is covered with a good grade of tissue paper or lightweight cloth. This covering is stretched tight and waterproofed with two or three coats of dope.

Experienced model-makers usually use gasoline engines to power their airplanes. Often they will build gas-powered models that they can control from the ground by working the plane's control surfaces, either by means of a control line or by radio control.

In many kits, modern "heat-shrink" plastics have replaced the traditional tissue and dope covering the airframe. Check with a hobby shop or model club to help you in selecting the type of kit you will build and fly.

Three general categories of powered flying models are available. These are free-flight, line-controlled, and radio-controlled model airplanes.

Free-Flight Models. As the term suggests, free-flight models are designed to be released with the engine running and to coast on oversized wings when the engine stops. Resembling sailplanes or gliders, these may stay aloft for an hour or more.

Control-Line Models. Also known as U-control models, control-line models are flown by two wires attached through one wing of the aircraft. These wires work the elevators on the horizontal stabilizer, controlling the altitude of the flight. The operator rotates inside a small circle, maneuvering the airplane up, down, or in loops.

Radio-Controlled Models. Many advanced model airplane builders install radios in their airplanes so the planes can be controlled in flight from the ground. A tiny receiver is built into the model, and signals are sent to it from a transmitter. These signals are used to guide the model's control surfaces, allowing the pilot to make the model dive and climb, turn, and cruise.

Model Power

There are three basic types of power plants for flying models: rubber-powered, fuel-powered engines, and battery-powered electric motors.

Engines for fuel-powered model planes are internal combustion engines that burn either an alcohol/oil mixture or gasoline. Most of them are two-cycle engines that draw fuel into the combustion chamber, fire, and expel exhaust by the up-and-down movement of the piston in the cylinder.

The simplest type is called the glow-plug engine. It is fired by a glowing filament heated by a battery. Once the engine starts, the heat of combustion keeps the glow plug hot. This engine uses a mixture of methanol, nitromethane, and oil as fuel.

The other common type of model engine is a simple diesel engine that burns an ether mixture and is fired on compression alone. The average flight time for a gasoline-powered engine is 10 to 15 minutes; although larger fuel tanks provide longer operation.

The battery-powered electric motors have an advantage; they can be switched on or off, thus providing maximum power instantaneously. Their disadvantages are a comparable lack of speed and shorter flight time.

Be sure you follow the safety precautions provided by the manufacturer for the fuel—or battery—operated model you build.

Before tackling requirement 3a, practice-build several of the simpler rubber-powered airplanes. Become familiar with common construction problems. Then, building a model with an engine will be less complicated.

Keep these tips in mind when building your model airplane.

- Be careful when using sharp tools such as razor blades and knives.
- Avoid inhaling fumes from airplane cement and airplane dope.
- Keep all materials and tools away from small children.

FLYING WITHOUT LEAVING THE GROUND

Flying Models in Competition

After you have built and flown several models, you may want to compete with other model flyers. In many communities, model clubs, hobby shops, civic organizations, and other groups sponsor meets. Some flying meets are held indoors in gymnasiums, armories, or other large halls.

There are events for gas- and rubber-powered free-flight models, gliders of various types, free-flight rocket-powered models, control-line models, and radio-controlled models of various kinds. These often involve "combat," during which the flyers' line-controlled, gas-engine models try to cut with their propellers streamers trailing from other models. Models are sometimes judged for workmanship as well as for flight.

National rules for model contests are established by the Academy of Model Aeronautics. This nonprofit organization's mission is to promote development of model aviation as a recognized sport and worthwhile recreation activity. See the resources section at the end of this pamphlet for contact information.

FLYING WITHOUT LEAVING THE GROUND

Flying Models Safely

Flying model aircraft requires a high level of skill and responsibility. The planes can weigh up to 100 pounds and reach speeds faster than 200 mph. Therefore, safety must never be compromised.

One commonsense rule of safety when flying model planes: Stay out of their way. Even a light, balsa-wood model powered by a rubber band might injure you if it hit you.

The Academy of Model Aeronautics publishes the National Model Aircraft Safety Code for use by AMA members and at AMA-sanctioned events, but it applies just as well to any model aircraft enthusiast. You can request a copy of the code (AMA document 105) by contacting the AMA or visiting the AMA Web site.

The Model FPG-9

An easy way to learn about how planes work is to build a model FPG-9. This simple design was devised by Jack Reynolds, a volunteer with the National Model Aviation Museum. The glider is made using a standard-size foam plate.

Materials Needed

- FPG-9 pattern
- 8⅞-inch foam plate
- Scissors
- Clear tape
- Ink pen
- Penny

In requirement 3b, you are asked to build an FPG-9 and compete with other Scouts in tests of precision flight and landing.

AVIATION 83

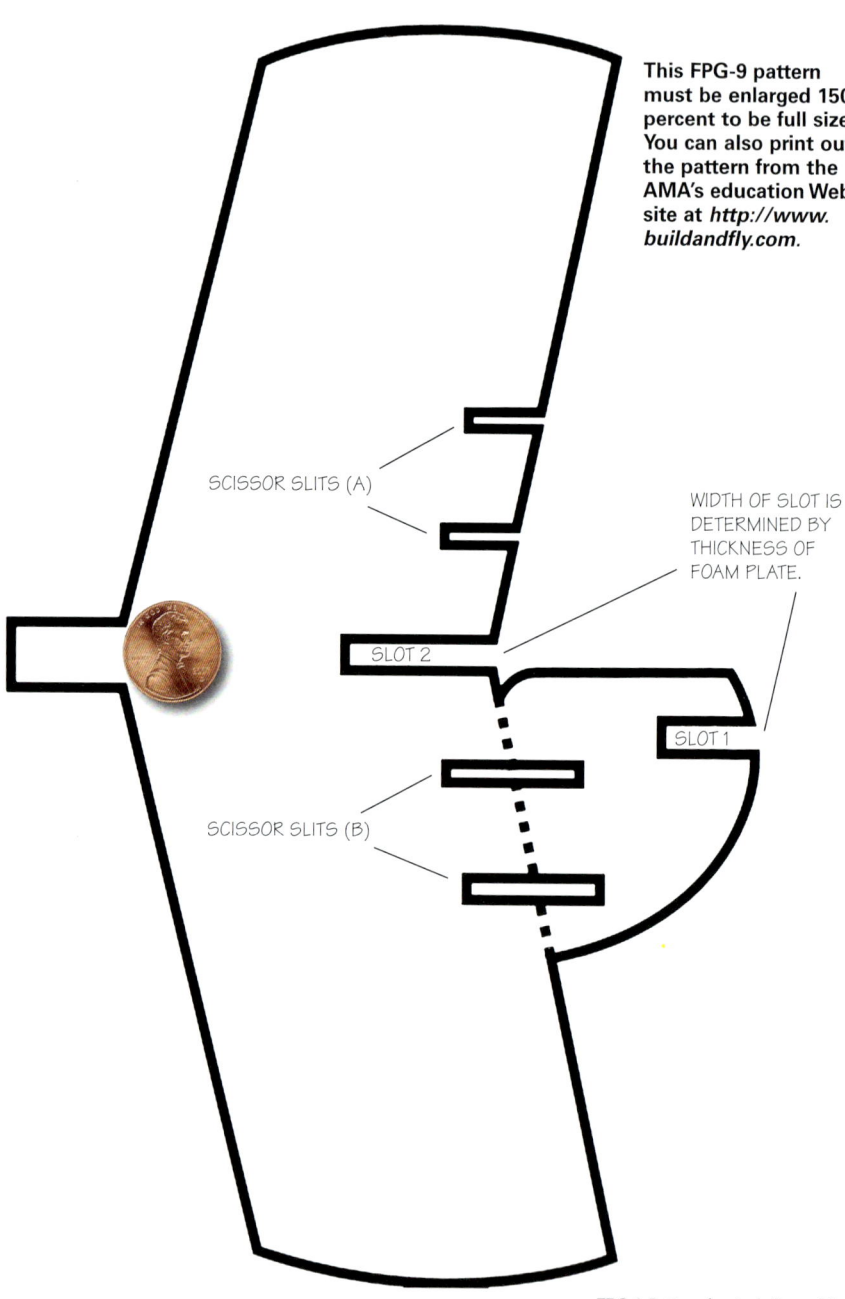

Making an FPG-9

Follow these steps to make your FPG-9.

Step 1—Photocopy the pattern provided (it will have to be enlarged 150 percent) and cut it out with scissors. Don't cut along the dotted line, just along the bold lines.

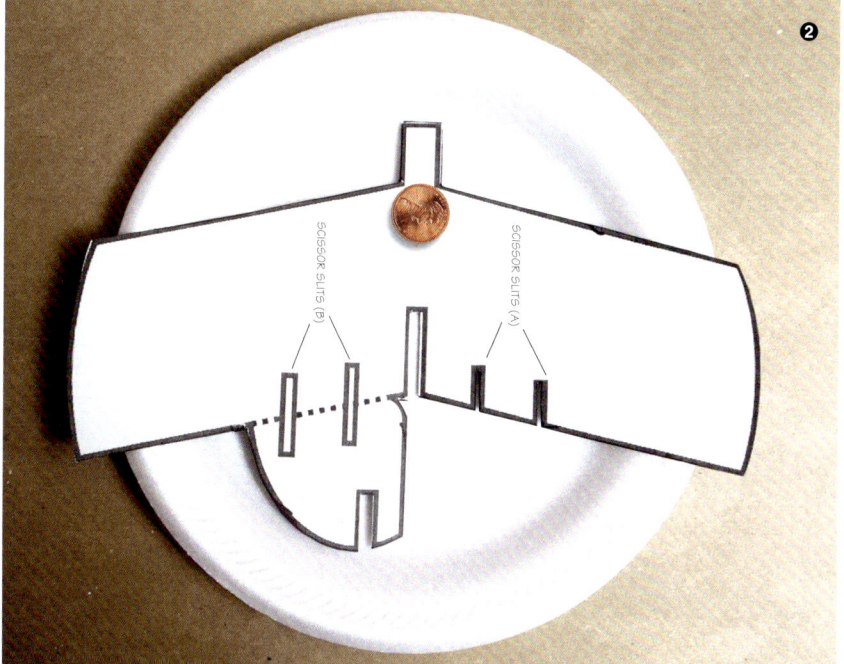

Step 2—Place the paper pattern in the center of the foam plate, making sure that the tail of the pattern stays on the plate's flat bottom, inside the curved portion. Don't worry if the tab at the top of the pattern is on the curved portion. The wingtips should spill over the curved edge of the plate.

Step 3—Trace around the pattern with the ink pen. Mark the scissor slots A and B.

Step 4—Cut out the foam template by following the pen lines you just drew.

Step 5—Cut along the dotted line to separate the tail from the wings. For ease, make all of your cuts from the outside of the plate toward the center of the plate. Don't try to turn the scissors to make sharp corners. Cut the two slots only as wide as the thickness of the foam plate; otherwise the pieces will not fit together snugly.

FLYING WITHOUT LEAVING THE GROUND

The FPG-9 uses "elevons" to control both pitch and roll. In a conventional airplane, elevators control pitch and ailerons control roll.

Step 6—To attach the tail to the wing, slide slot 1 into slot 2. Making sure the tail is perpendicular to the wings, use two small pieces of tape (about 2 inches long) to secure the bottom of the tail to the bottom of the wing.

Step 7—Glue the penny near the front of the plane where indicated on the pattern. Fold the square tab back over the penny; tape it down to secure the coin.

AVIATION

Step 8—Bend the elevons (the flaps on the back edges of the wings) upward. This will provide for a flatter glide path.

Step 9—Your FPG-9 is ready to fly. Gently toss the plane directly in front of you. Once it flies reasonably straight and glides well, try throwing it hard with the nose of the glider pointed 30 degrees above the horizon. The glider should make a big loop and have enough speed to glide 20 to 25 feet after completing the loop.

FLYING WITHOUT LEAVING THE GROUND

Testing Your FPG-9

Now that you have built your FPG-9, you can test the effect of various modifications. See what happens when one elevon is up and the other is down. Make both elevons neutral (even with the wing) and move the rudder (the flap on the tail) to the left. What happens? Figure out how to make the glider fly to the right.

Once you are an FPG-9 ace, challenge other members of your troop or patrol to make their own gliders. Then, have a competition to test your gliders. To judge precision flight, hang a plastic hoop about 20 feet from the starting line and try to fly your gliders through it. To judge precision landing, lay the hoop on the ground and try to land your gliders within its circle. Give each participant a set number of tries and number of points per successful try, then total each person's score.

FPG-9 design courtesy of Jack Reynolds.

AVIATION

Careers in Aviation

Aviation offers a nearly unlimited variety of career opportunities—many of which do not directly involve airplanes. Experts in the field estimate that for every person who flies an aircraft, there are 600 others who fill aviation-related positions. The Federal Aviation Administration has established seven categories of aviation employment.

- Pilots and flight engineers
- Flight attendants
- Airline nonflying careers
- Aircraft manufacturing
- Maintenance/avionics
- Airport careers
- Government careers

Nearly all workers in the aerospace industry must be highly skilled. Those involved in the manufacture, flying, and maintenance of aircraft and spacecraft are especially well-trained. The quality of study and work that might give you just a passing grade will not be good enough for a position in aviation.

> As you consider whether you want to enter the aviation field, ask yourself about your skills and interests. Are you mechanically or scientifically minded? Do you work well with your hands? Do you work well with people? If you have the necessary abilities, there may be an opportunity for you in aviation—provided you are willing to work hard to learn.

CAREERS IN AVIATION

For more information about the resources mentioned in this section, see the resources section at the end of this pamphlet.

If you are serious about a career in aviation, you should begin planning for it when you enter high school because you may need a background in such fields as mathematics and physics. For many positions, college degrees are necessary, again usually with emphasis on math and science. Depending on the career you want to pursue, you may choose a college that offers a specialized aviation program. In addition, many commercial pilots earn their wings as members of the U.S. armed forces. Your guidance counselor can help you learn more about how to prepare for an aviation career.

The FAA offers several publications detailing aviation career areas. For information, visit the FAA Web site, or write to Superintendent of Documents, Retail Distribution Division, Consigned Branch, 8610 Cherry Lane, Laurel, MD 20707. For more information on aviation-related university studies, contact the University Aviation Association.

This chart shows school subjects that must be mastered for some of the many careers in aviation.

CAREERS IN AVIATION

AVIATION CAREER	Chemistry	Physics	Bookkeeping	Business Machines	Biology	English/Technical Writing	Mathematics	Speech	Mechanics	Sheet Metal	Machine Shop	Welding	Electronics	Electricity	Mechanical Drawing	Composite Materials	Journalism	Psychology	Keyboarding	Computers	Photography	Nutrition	Meteorology	Geography
Aerial Photographer	•	•				•		•									•		•	•	•		•	•
Aeronautical Engineer	•	•				•	•	•	•						•	•			•	•				
Air Electronics Officer		•				•	•	•					•	•					•	•			•	
Aircraft Manufacturer						•	•	•								•			•	•				
Aircraft Mechanic		•				•	•	•	•	•	•	•		•					•	•				
Aircraft Navigator						•	•	•						•	•				•	•			•	•
Aircraft Sales						•	•	•											•	•				•
Airframe Mechanic		•				•	•	•	•	•	•	•			•				•	•			•	•
Airline Accountant			•	•		•	•	•											•	•				
Airline Maintenance Inspector	•	•				•	•	•	•	•	•	•	•	•	•				•	•			•	•
Airline Pilot	•	•				•	•	•	•				•	•					•	•			•	•
Airline Reservations						•	•	•										•	•					•
Airline Scheduler/Dispatcher			•	•		•	•	•											•	•				
Air Traffic Controller						•	•	•					•	•				•	•	•			•	•
Airport Management						•	•	•											•	•			•	•
Avionics Technician		•				•	•	•					•	•					•	•				
Chemical Engineer	•	•			•	•	•	•								•			•	•				
Corporate Aviation Manager			•	•		•	•											•	•	•			•	•
Corporate Pilot		•				•	•	•	•				•	•					•	•			•	•
Customer Service Rep.						•	•	•										•	•	•				
Design Draftsman						•	•	•							•	•			•	•				
Electronics Engineer	•	•				•	•	•	•				•	•	•				•	•				
Fixed Base Operator			•	•		•	•	•										•	•	•			•	•
Flight Attendant						•	•	•										•	•	•		•		•
Flight Engineer	•	•				•	•	•	•				•	•					•	•			•	•
Flight Surgeon	•	•			•	•	•	•										•	•	•		•		
Flight Test Mechanic		•				•	•	•						•					•	•			•	•
Instrument Mechanic	•	•				•	•	•					•	•					•	•				
Line Service Specialist			•	•		•	•	•										•	•	•				
Mathematician		•				•	•	•											•	•				
Mechanical Engineer	•	•				•	•	•							•	•	•		•	•				
Metallurgist	•	•				•	•	•		•		•							•	•				
Meteorologist	•	•				•	•	•					•	•					•	•				•
Military Pilot	•	•				•	•	•					•	•					•	•			•	•
Physicist	•	•		•		•	•	•						•					•	•				
Public Relations						•	•	•									•	•	•	•			•	•

Aviation Resources

Scouting Literature

Auto Mechanics, *Electronics*, *Engineering*, *Model Design and Building*, *Orienteering*, and *Space Exploration* merit badge pamphlets

> For more information about Scouting-related resources, visit the BSA's online retail catalog (with your parent's permission) at *http://www.scoutstuff.org.*

Books

Echaore-McDavid, Susan. *Career Opportunities in Aviation and the Aerospace Industry: A Guide to 80 Careers in Aviation and the Aerospace Industry.* Checkmark Books, 2005.

Eichenberger, Jerry A. *Your Pilot's License.* McGraw Hill, 1999.

Goldstein, Avram. *The Right Seat: An Introduction for Would-Be Pilots.* AirGuide Publications, 1996.

Gunston, Bill. *Aviation: The First 100 Years.* Barron's, 2002.

Lopez, Donald S. *Aviation: A Smithsonian Guide.* Macmillan, 1995.

Nahum, Andrew. *Flying Machine.* DK Publishers, 2004.

Rabinowitz, Harold. *Conquer the Sky: Great Moments in Aviation.* Metro Books, 1996.

Periodicals

Flying
HFM Inc.
Telephone: 212-767-6000
Web site: *http://www.flyingmagazine.com*

Model Airplane News
Air Age Media
Toll-free telephone: 800-877-5169
Web site: *http://www.modelairplanenews.com*

Plane & Pilot
Werner Publishing Corp.
Telephone: 310-820-1500
Web site: *http://www.planeandpilotmag.com*

Organizations and Web Sites

Academy of Model Aeronautics
5161 East Memorial Drive
Muncie, IN 47302
Toll-free telephone: 800-435-9262
Web site: *http://www.modelaircraft.org*

AVIATION RESOURCES

ALLSTAR Network
Aeronautics Learning Laboratory for Science, Technology, and Research
Web site: *http://www.allstar.fiu.edu*

The Aviation History Online Museum
Web site: *http://www.aviation-history.com*

AvStop Magazine Online
Web site: *http://www.avstop.com*

Chasing the Sun
The History of Commercial Aviation Seen through the Eyes of Its Innovators
Web site: *http://www.pbs.org/kcet/chasingthesun*

Federal Aviation Administration
800 Independence Ave. SW
Washington, DC 20591
Toll-free telephone: 866-835-5322
Web site: *http://www.faa.gov*

Flight-History.com
Web site: *http://www.flight-history.com*

National Aeronautics and Space Administration
Suite 1M32
Washington, DC 20546-0001
Telephone: 202-358-0001
Web site: *http://www.nasa.gov*

National Oceanic and Atmospheric Administration
14th Street and Constitution Avenue, NW
Room 6217
Washington, DC 20230
Telephone: 202-482-6090
Web site: *http://www.noaa.gov*

University Aviation Association
3410 Skyway Drive
Auburn, AL 36830-6444
Telephone: 334-844-2434
Web site: *http://uaa.auburn.edu*

Videos and DVDs

Sporty's Airships. VHS, 50 min.

Larry Bartlett's Flying the Alaska Highway. VHS, 75 min.

Wonderful World of Flying's Have Plane, Will Travel. VHS, 70 min.

Sporty's Learning to Fly. VHS, 55 min.

Sporty's Non-Flier's Guide to the Cockpit. VHS, 75 min.

Sporty's So You Want to Fly Helicopters? VHS, 110 min.

Special Helps

Aviation Merit Badge Kit, Cessna Aircraft Company, Air Age Education, Attention: Joyce McKenzie, Dept. 407, P.O. Box 1996, Independence, Kansas, 67301, $1.50. The kit contains materials and information to help you earn this merit badge. In addition, a plastic airplane model that can be used to demonstrate the aircraft control surfaces is available for a nominal charge.

Acknowledgments

The Boy Scouts of America gives special thanks to Sporty's Pilot Shop for making this major revision of the *Aviation* merit badge pamphlet possible. In particular, thanks to Mark Wiesenhahn for his assistance.

Thanks to Robert Lamb Jr. for his work on the manuscript for the 2000 edition of this pamphlet, upon which the new edition is based. The BSA thanks Ed Mitchell and his colleagues in Phantom Works at Boeing Aircraft Company for their assistance with the 2000 edition, as well.

We appreciate the Quicklist Consulting Committee of the Association for Library Service to Children, a division of the American Library Association, for its assistance with updating the resources section of this merit badge pamphlet.

The BSA is grateful to Rear Adm. J. Dan McCarthy, U.S. Pacific Fleet, U.S. Navy, for facilitating the use of U.S. Navy photographs in the production of this book.

Thanks also to staff members at the Academy of Model Aeronautics for their time and expertise in reviewing the text, and for providing some of the photos used in this pamphlet.

Photo and Illustration Credits

Academy of Model Aeronautics, courtesy—pages 79 *(top)* and 82

Academy of Model Aeronautics/ Jack Reynolds, courtesy—page 84

©2005 JupiterImages Corporation— pages 36 *(background)* and 49 *(inset)*

Library of Congress, Manuscript Division, Washington, D.C., courtesy—page 7

NASA, courtesy—page 42

National Oceanic and Atmospheric Administration, courtesy— pages 9, 46 *(top)*, and 57

©Photos.com—cover *(all planes, control tower)*; pages 3–6 *(all)*, 13 *(background)*, 16, 18 *(photo)*, 20 *(both)*, 31, 35 *(top)*, 39 *(bottom)*, 40 *(background)*, 43 *(background)*, 52, 59, 61 *(both)*, 66–67 *(all)*, 69, 70 *(right)*, 74, 76 *(background)*, 79 *(background)*, 80, 81 *(background)*, 90, and 91–92 *(both)*

©Pratt & Whitney, courtesy—page 30

©U.S. Air Force, courtesy—page 8

U.S. Department of Agriculture/Tim McCabe, courtesy—page 13 *(inset)*

U.S. Department of Agriculture/Bob Nichols, courtesy—page 12

Wikipedia.org, courtesy—cover *(wind indicator)*; page 50

All other photos and illustrations not mentioned above are the property of or are protected by the Boy Scouts of America.

Daniel Giles—pages 40 *(inset)*, 60 *(bottom)*, and 71

John McDearmon—14–15 *(all)*, 17 *(all)*, 18 *(illustrations)*, 21–23 *(all)*, 25, 27–28 *(all)*, 51, and 64

Toby Everett—page 11

Brian Payne—pages 10, 70 *(left)*, and 72

Randy Piland—pages 68 and 73